ONE MOON, ONE CIVILIZATION

WHY THE MOON TELLS US WE ARE

ALONE IN THE UNIVERSE

ONE MOON, ONE CIVILIZATION

WHY THE MOON

TELLS US WE ARE

ALONE IN THE UNIVERSE

BY

JORGE LABORDA

First Edition. March, 2011

© Jorge Laborda, 2011

ISBN # 978-1-4467-1245-0

Editor and Publisher: Jorge Laborda

Cover: Jorge Laborda

Printed by Lulu (www.lulu.com)

Dedication

I wish to dedicate this book to the memory of Galileo Galilei. There are good reasons for this. First, in 2010, we celebrated the 400th anniversary since this universal scientist, in January 1610, aimed his telescope at Jupiter and discovered that there were other moons besides our own and, therefore, that not everything in the universe revolved around the Earth, or even around the Sun. in March that year (the same month I finished writing the initial version of the book you are holding in your hands, or reading on a computer screen), Galileo published, in Latin, his short treatise entitled "Siderius Nuncius" (which translates as Sidereal Messenger), in which he described the observations of the moon, stars and satellites of Jupiter made with his newly invented telescope. It is, therefore, the first description of astronomical observations for which an instrument, a product of the technological progress in another area of science, in this case, optics, was used. Since then, the progress of astronomy always has gone hand in hand with advances in technology, and it still happens today.

However, the main reason, in my opinion, for which Galileo deserves mine and many more dedications in his memory is that he was one of the few brave men who taught us that we should not be afraid of knowledge, even when it can endanger our very existence or, I add, jeopardize what we believe our existence means, its own transcendence in the material universe, or beyond. Reality is one. Neglecting its knowledge, trying to distort it, or rejecting it because it is contrary to our preconceptions only leads to a halt in our progress as human beings, but it will never change it. On the contrary, we will only have the opportunity to change reality if we know it, as clearly demonstrated by the technological advances we enjoy, enabled by the increasing scientific knowledge, that have changed our reality. For his courage in accepting the reality in front of all those who refused it; for his determination to embrace the light of scientific knowledge against religious obscurantism, I dedicate this book to Galileo.

About the author

Jorge Laborda graduated in Chemistry in 1981. He completed the work of his Ph.D. doctoral thesis at The Center for Scientific Research on Cancer, Paris, France, where he worked from 1984 to 1987 and defended his Ph.D. at the University of Zaragoza, Spain, in 1987. He moved that year to the National Institutes of Health in Bethesda, MD., for a postdoctoral stay of one year. The following year, he joined the Lombardi Cancer Center, at Georgetown University, Washington DC, where he worked for three years. In 1991, he joined the Center for Biologics Evaluation and Research (CBER), Food and Drug Administration (FDA), located at the National Institutes of Health campus, USA, where he became a director of research and got a tenured position as Principal Investigator. In 1999, he joined the Medical School, University of Castilla-La Mancha, as Head of the Biochemistry and Molecular Biology area. Dr. Laborda has worked in nearly forty research projects. He has been responsible for the review of numerous projects on new cancer therapeutics proposed by European or American biotechnology companies. It has also been the president of the American committee for drafting regulations on the use of transgenic plants for production of pharmaceuticals for human use. The result of his research is the discovery of two genes involved in the control of cell differentiation and his important contribution to the discovery of the receptor for an important differentiating factor. From November 2003 to May 2004, Dr. Laborda joined the European Commission as a National Expert to work in the management and promotion of pioneering biomedical research areas, such as synthetic biology. In April 2004, Dr. Laborda was elected

Dean of the Medical School UCLM, a position he held until April 2008.

Dr. Laborda is also a popular science writer and journalist. He published the books "The Gods have been cloned", "The Arabian bases of DNA and other scientific stories", "The funnel of intelligence and other assays" and "Adenio Fidelio", a science-based novel for teens. He has participated numerous times in Spanish international radio broadcasts, and local television dedicated to science, and is an occasional columnist in the journals El Pais, the Heraldo de Aragón and weakly columnist in La Tribuna de Albacete. The results of this work are more than 500 popular science articles published to date. Currently Dr. Laborda holds the position of Counsellor of Consumer Affairs, Science and Technology of the City of Albacete. Spain.

Acknowledgments

This book, as almost any book written by any author, would not have been possible without the help, advice or patience of other people. I would like to thank, in particular, Dr. Carlos Elías, Dr. Alberto Nájera, Dr. Fernando Cuartero, Dr. Pascual González, and Dr. Enrique Díez for their comments, which helped to improve it. I especially thank my wife, Rosa, for her endless patience with my cosmic absentmindedness and my lunatic mood swings.

Table of Contents

Foreword

The question whether or not we are alone in the universe continues to be one of the most intriguing enigmas of science. Surprisingly, this mystery did not exist when modern science started to walk its first and hesitant steps, facing the opposition of the religious ideas from the Judeo-Christian tradition. At the beginnings of the XVIIth century, when Galileo defended Copernicus' ideas, very few doubted that humankind was the king of divine creation and, consequently, that we, humans, were alone on the peak of creation.

As with so many other mysteries, this one is of the kind that, paradoxically, becomes deeper when science has already increased significantly the knowledge about the world. The development of better telescopes that the one invented by three Dutch scientists[1], and improved substantially by Galileo[2], leaded to the revelation that our galaxy is formed by billions of stars, many of them similar to our Sun. This opened up the door to the idea that solar systems similar to ours could harbor life, and even intelligence.

Even in the second decade of the XXth century, it was still not clear whether the universe was bigger than our galaxy, the Milky Way. However, thanks to the development of yet more potent telescopes, it could be demonstrated that our universe is much bigger that our galaxy and it contains billions of galaxies similar to ours.

All these discoveries suggested that the enormity of our universe could not have the only purpose of housing humanity alone as the maximum universal exponent of life and intelligence (which would be pathetic, to say the least). In words of the main character of the movie *Contact*: "it would be an awful waste of space". One was almost forced to draw the conclusion that "somebody" must be out there. It was only a question of time to find it.

These ideas facilitated the development of a great deal of science-fiction stories and movies, in which, repeatedly, the humanity contacted, or collided, with extraterrestrial civilizations. I believe the pioneer work of this gender and topic was *The War of the Worlds*, a novel written in 1898 by Herbert George Wells (better known as H.G. Wells)[3], which has been brought to the movies in several occasions, one of them recent. In 1938, the radio broadcast of a version of this story by Orson Welles, who later became a famous movie director, spread the panic among many radio listeners, who literally believed that planet Earth was being invaded by a military force coming from Mars[4]. There is no doubt that, besides the realism and quality of the broadcast, if most listeners had not believed, in those early years, that intelligent life was possible on other planets, the panic caused by the broadcast would not had taken place.

From then on, we have been continuously accompanied by the most varied and extraordinary extraterrestrial beings in the literature, the cinema and, according to some, unbeknownst to us, even in real life. For example, who of a certain age doesn't recall the popular TV series of the 60's, *The Invaders*?[5] In this case aliens were very similar to humans, except they possessed a rigid little finger. Can you come up with a more absurd difference between humans and aliens? If not more absurd, surely we can make it more amusing by simply imagining that the difference may reside in the rigidity of a much more notorious masculine appendix than the little finger! Nevertheless, this would prevent identifying women as aliens, which, at the current time, is not politically correct, not even for female aliens.

Of course, let us don't forget the double trilogy (I don't dare to call it sexology, since it has very little sexy quality) "*Star Wars*[6]", in which we encounter the most extraordinary and strange alien races of all physical or psychic conditions. Fortunately, none of the different alien races possesses a stiff little finger, or any other stiff finger for that matter, although they can possess extra long tongues, or exhume thick liquids as those secreted by snails and slugs.

And what about the UFO phenomenon? Many humans consider these Unidentified Flying Objects as the factual evidence proving we are being visited and observed by alien intelligences. Intelligences which are sufficiently stupid, however, as to only appear in front of some lost farmers from Missouri (USA) or Patagonia (Argentina), and doing so after a many light-year long trip, which is to say, a very, but very, very, very long trip. If aliens intended establishing contact with a not so intelligent human being and, through him or her, with the entire human race, couldn't they contact the President of the United States, who usually is not anyway a lot smarter than any farmer from Missouri? What an awful loss of time if they don't do it soon!

Now, seriously, it appears that the idea that we are not alone in the universe is accepted and very widespread among thinking humans, and even more among those not thinking much. It is an idea that provides us with a certain relief. We do not like the feeling of being alone, not even when the company can become a threat.

And talking about threats, it is remarkable that scientists with the prestige of Stephen Hawking, recently (April, 2010) has brought the alien subject up to become one of the current affairs in the press. Hawking, in his new TV series *Into the Universe with Stephen Hawking*, warned us about the danger that will entail responding to any alien civilization trying to contact us, thus confirming our space coordinates so they can come, visit ... and conquer us! Many scientists, certainly Hawking does, believe the universe is probably full of civilizations more advanced than ours, which can contact with us at any moment.

Nevertheless, others support the idea than Earth is a rarity in the universe. This idea has been called the Rare Earth Hypothesis, and was put forward in the book *Rare Earth: Why Complex Life Is Uncommon in the Universe*, written by Peter Ward, a geologist and paleontologist, and Donald E. Brownlee, an astronomer and astrobiologist[7]. In their book, the authors analyze the factors that are required so that complex life may arise on a given planet, and they conclude that there are so many and they are so unlikely that there mustn't be many planets like ours in the entire universe.

However, recent astronomical discoveries demonstrate, indeed, that we are not alone in the universe, at least in what refers to planets and planetary systems. Hundreds of extra solar planetary systems have been recently discovered[8]. From the study of their frequency and distribution, the conclusion can be drawn that maybe a trillion trillion planets exist in the observable universe. Nevertheless, these discoveries also indicate that our Solar System is a rarity among the other solar systems, which usually have giant planets, similar or bigger than our Jupiter, orbiting the central star even much more closely than Mercury orbits the Sun. This rarity is not a rarity without importance; rather it could have affected in a fundamental way whether or not it was possible that in one of the other orbiting planets, in our case our dear planet Earth, life arose and had the time and the relative tranquility to evolve and develop a technological civilization.

The rarity of our Solar System, with its small and rocky planets close to the central star and the bigger planets orbiting far from it, is not only limited to the planetary distribution; it includes also our own planet Earth and its satellite, the Moon. I can assure you the Moon is a very strange satellite and for that reason its presence beside Earth might have been decisive to be here now writing about it. Let make it clear, I do not mean the Moon has been essential for life arising on Earth. What I claim is that the Moon has been decisive so that we have acquired enough intelligence and technology as to be able to discover its important influence on life and civilization on Earth.

In this book, we will speak about all this, because the accumulated scientific knowledge, in spite of having demonstrated that the universe contains billions of galaxies, and probably more than a trillion trillion planets, suggest that, anyway, it is still likely that, for practical purposes, we are alone in the universe and, indeed, that this results in an awful waste of space. Below, I will try to explain whether this can be true, and why.

We will start our journey from the origin of life itself, and we will explain why it is not likely that life based on a different chemistry than ours can exist at any other place of the entire universe. We will continue our journey by establishing connections among scientific knowledge in diverse areas of science until reaching a conclusion, a thesis, of which the reader will judge its validity. In summary, we will make a small journey along the science of life, the universe and intelligence. I hope the journey, although it will drive us nowhere, as it happens to most of the journeys we make in life, will allow us to enjoy it and learn something new, which, in my opinion, is always the most important goal of any journey, included the very journey of human life.

Do not forget to buckle up, because off we go.

Jorge Laborda. December, 2010.

Notes to foreword

1 http://galileo.rice.edu/sci/instruments/telescope.html

2
http://www.archive.org/stream/galileohislifean011377mbp/galileohislifean011377mbp_djvu.txt

3 http://www.s4ulanguages.com/hgwells.html

4 http://en.wikipedia.org/wiki/Orson_Welles

5 http://en.wikipedia.org/wiki/The_Invaders -
http://www.youtube.com/watch?v=g3fu3iIfRK4

6 http://en.wikipedia.org/wiki/Star_Wars

[7] Peter Ward and Donald E. Brownlee (2000). "Rare Earth. Why complex life is uncommon in the universe" Ed. Springer (December 10, 2003). ISBN-10: 0387952896. ISBN-13: 978-0387952895

8 http://exoplanet.eu/

Chapter 1: Universes

One of the most interesting philosophical problems, in my view, is why there is something rather than nothing. Don't worry; I am not intending to resolve this issue here, even if science provides us with some partial answers to that question. Besides, even more interesting is the problem of why there is a being self-aware of its own existence, rather than only beings ignoring it. In other words, why do we exist, instead of only inanimate objects or life forms devoid of consciousness? Any reason? Who or what decides what must exist and what must not?

Again, I am not intending to deal here with this problem. You may continue reading in peace! But the indubitable fact is that we inhabit a universe in which life is possible. Besides, not only are we in a bio-permissible universe, but in a universe that, in addition to allowing the origin and evolution of life, has allowed the origin and evolution of intelligence, consciousness, and, at least, one technologically advanced civilization, capable of communicating with the use of electromagnetic waves: ours. Why it is this way, rather than any other?

Likely, those who hold some kind of religious beliefs would answer those questions by appealing to them: the universe was created by God; designed by Him. We are the making of His creative ingenuity, and here we are. Period. The mystery is not this world, but what await us in the next.

But even if these ideas are true, we can still ask why there is a God rather than nothing, and why there is a God knowing

He/She exists, rather than a God not knowing it (even if, obviously, this ignorant God would not be "God-looking"). To get answers to these questions, especially to the questions about our own existence, which is what we really need explaining and give meaning to, science provides us today with many data and facts that we need to consider. Of course, these new knowledge cannot provide answers to the fundamental questions yet, but lets not make the mistake of believing that because science cannot answer those fundamental questions now, it could never do it. Scientific research has been, and still is, the endeavor that better responses about the world, about our condition as human beings, and about the reason for our passions, needs, and character has provided and continues to provide. If science cannot fully answer yet questions as those posed above, maybe it could do it one day. Perhaps that day, finally, we will talk to God, if He/She exists, or otherwise discover that we are alone; that we are the simple result of the behavior of matter, matter that exists because, necessarily, either something has existed forever or it has started to exist at a given "moment", even if this "moment" must be placed necessarily outside space and time, because even them did not exist.

Let's set aside these questions, for the time being, and enter into better-known territories. As I said, it is obvious that our universe is of such a nature as to allow the development of life. Science knows nowadays that both, the conditions of the initial explosion that apparently gave origin to the universe (the Big Bang), and the very laws of Nature governing the behavior of matter, cannot be very different from what they are so that life can develop. For instance, suppose that the force of the initial explosion were greater than it really was, perhaps a 10% or 20% greater. In this case, it is possible that matter would have dispersed too quickly to allow its assembly into stars and galaxies. If stars did not form, life would be impossible. On the contrary, let's suppose that the power of the initial explosion were less than it really was. In this situation, it is possible that matter would have collapsed back to its initial state before the necessary time had elapsed so that life, and of course intelligence, could develop.

We can imagine similar types of thought experiments not only with the initial power of the explosion, but with the very laws governing the behavior of matter, space, and time. For instance, if the force of gravity had been weaker than it is, the universe would have spread also very quickly, and stars could not have formed. If gravity had been stronger, the power of the initial explosion being equal, matter would have gathered again very quickly, not allowing a sufficient time for the development of complex life.

We find a similar situation with the other forces operating in the universe: the strong nuclear force, the electromagnetic force and the weak nuclear force. These latter two, even if they appear today as different forces in Nature, were united in only one force when the universe was born[1]. If the strong force were weaker than it is, many chemical elements, some essential to life, as we know it, could not exist. The strong nuclear force holds the protons, positively charged, together in the atomic nucleus, despite the electrostatic repulsion that they exert on each other. As we know, the number of protons in the nucleus determines the chemical nature of the elements. For example, carbon has six protons in the nucleus;

iron, twenty-six. If the strong force were perhaps only a little weaker than it is, iron could either not be formed or be unstable and disintegrate, as it happens now to most heavy radioactive elements, such as uranium or plutonium. In a universe lacking iron, no doubt life as we know it would be impossible.

Thus, the laws and constants of Nature cannot be arbitrarily different from what they are or, otherwise, life could not develop in our universe. There is, therefore, a bio-permissible range of values for the forces acting on matter. If these values were different in some undetermined, but small, degree, we would not be here discussing about these critical issues.

It is, no doubt, astonishing that we, beings generated by the universe we inhabit, can now imagine other universes that would behave according to different natural laws and would or would not generate in them other beings endowed with imagination. What this seems to suggest is that it appears unnecessary that we are here. There may be a multitude of different universes, in most of which life would not exist. There could be also a multitude of different universes harboring life, but most of them devoid of intelligent beings. According to the laws of probability and statistics, and considering the rather narrow range of possible values for the forces and constants governing the evolution of matter that are compatible with the existence of life, it is clear that the universes in which life and intelligence could develop would be a minority among the possible universes, if other universes existed (a theory, called the multiverse theory, seriously considered by some scientists. We will talk about it in more detail later). However, despite all probabilistic considerations are against us, here we are.

Having determined that the universe in which we live is suitable for the development of life and civilization, unlikely as it might be, we may ask: would they necessarily develop? In other words, in a universe fit for life, would it always develop? In a universe suitable for intelligence and civilization, would they always develop? Intuition tells us that it must not be necessarily so. We can imagine universes like ours in which, however, Earth and its inhabitants did not exist, nor intelligent beings on other planets. The universe would be perfectly well without them, even better than with them, considering how things are going lately on our planet. Therefore, life and intelligence would not necessarily arise even in a universe fit for them, and whether this happens or not may depend on random or chaotic factors, perhaps on how the initial "explosion" that caused everything developed, despite the laws of Nature governing those other universes were identical to ours.

Chance or necessity

However, to be intellectually honest about this issue (and we should at least be intellectually honest if we can not be it otherwise) it is necessary to consider the factor of chance. Does chance exist or not in our universe? This is a question that very few philosophers and scientists have left to chance alone. Undoubtedly, there are two possible and clearly defined positions on this issue: chance exists or it does not. The intermediate position, chance exists and does not exist at the same time, is too chancy to consider it seriously.

Now seriously, according to my knowledge, current Physics maintains that randomness is an intrinsic component of the universe[2]. This means that in situations in which all the causal variables are defined, the result may be still uncertain. Quantum mechanics uses the concept of probability to predict the behavior of particles and atoms, and its success as a scientific theory seems to agree with those who hold a position in favor of chance. Similarly, the phenomenon of radioactivity suggests that chance exists. In a population of radioactive uranium atoms, for example, we can calculate exactly how many of these atoms will have decayed in a period of time, but we cannot predict when a particular atom will do it. It could do it immediately, or take thousands of years, or even never disintegrate. Which specific atoms would decay at any given time seems, therefore, a matter of chance, despite it is determined, for each radioactive element, the number of them that will decay in a given period of time[3].

However, as mentioned above, it is possible to consider an alternative position, very well summarized by the words of Einstein: "*God does not play dice*" (to which, it seems, the great physicist Niels Bohr answered: "*Einstein, stop telling God what to do*"[4]). This position is as follows: if from the first moment of the universe's life everything follows certain laws, the laws of Nature, which don't vary with time, then everything is predetermined from its origin, and will occur according to the initial conditions and to the application of the laws along time. Chance would be but an illusion due to our inability to know everything. For example, in the case of radioactive decay, we cannot know for sure which atoms will disintegrate next second, but that does not mean they are not predetermined to do so. Indeed, the fact that we can predict the proportion of those atoms that will disintegrate in a given time can be interpreted as it is predetermined that some of them will do it, although we don't know how to identify which ones and why it happens to be them in particular, but not others. So, we could conclude that it is precisely because it is certain and determined that some particular atoms will disintegrate in a given period of time, that we can predict how many of them will do it in average and deduce the laws of radioactive decay.

A simple example on how the initial conditions and the obedience to some particular rules determine the final state is the Sudoku game. The initial state of a Sudoku shows particular numbers whose positions in the squares determine the positions of the unknown numbers that we have to discover and place in the appropriate empty squares, by applying the rules proper of Sudoku, and the laws of logic. The fact that initially we don't know what number should be placed in a particular square doesn't mean this number must not be precisely one and not another, so that the Sudoku is correctly solved.

The classification of an event as random, for example, the rolling of a die (if God is not rolling it), is determined by the impossibility to foresee the outcome. However, this does not necessarily mean the outcome is not predetermined. Unforeseeable means that the human mind is unable to foresee the outcome of a future event, not even with the help of the best computers, despite which the outcome could be predetermined. Indetermination, on the other hand, means that the outcome of an event was not previously fixed and it could have happened in a different way from which it had. Returning to the example of rolling a die, for instance, if the outcome is a five we can think that this was not predestined to be, and that a different outcome was equally possible. However, a problem with this way of thinking is that *each event in the universe is unique in the space-time continuum, and it can never be repeated.* In other words, if when rolling a die playing Parcheesi the outcome is a five, and when rolling it again *later*, the outcome is an ace, we cannot conclude in a definitive way that those outcomes were not predetermined from the beginning of the universe. This is so because each one of the outcomes happens in a different space-time moment of the evolution of the universe and, therefore, not under identical conditions. Between the instant of the first rolling and that of the second, the universe has expanded, the Sun has transformed mass into energy and it exerts a slightly smaller gravitational pull on Earth and on the die, the Moon has advanced in its orbit and its gravitational attraction has also changed... We could continue this way on and on, listing the conditions which were different between the instant of the first rolling and that of the second, but to list them all would require more time than the universe has left until it dies, and we don't have that much.

At any rate, the laws of Physics are compatible with the fact that the die may fall once on a side, and later on another or the same side. That is to say, in a rolling series, the outcome will be near, but almost never equal, to that 1/6 of the times the die falls on each number. However, this is still compatible with the possibility that each individual outcome, happening, I repeat, *only a single time* in the space-time, is predetermined.

In other words, to prove that an event happens at random it would be necessary to repeat it in the same instant of the space-time evolution of the universe. That is, to demonstrate that chance exists, after rolling a die at 11:00 o'clock and getting a four, we would have to go back in time, precisely until 11:00 o'clock again, roll the same die and get a different outcome this time, let say, a three. Then, we could return to 11:00 o'clock again, and roll the die once more. This time we may get a two. If something like this happened, we would have demonstrated that chance exists and that the same event happening in the same space-time conditions of our universe can yield alternative outcomes.

If, on the contrary, when returning back in time until 11:00 o'clock and rolling the die we again get a four, and after returning back in time another three or four times and repeating the rolling, the outcomes were always four, we could suspect that chance doesn't exist in the universe, although we could not conclude this for sure. In fact, this result would be paradoxical, for it would be probabilistic in nature, because we must go back in time, and roll the die, infinite times and always get a four to prove that chance does not exist. As it would be impossible to do that, even if we were able to go back in time, we would have to be satisfied to conclude, only with a greater or smaller probability (depending on the number of times that we have repeated the experiment), that chance does not exist. However, importantly in this case, this probability would be a consequence of our inability to know for sure, not to the existence of randomness.

In any case, these experiments are obviously impossible to perform. Therefore, we cannot prove experimentally whether or not chance exists in the universe, neither can we infer its existence from the outcome of "random" events in time, since each can be individually determined from the origin of the universe due to the immutability of the laws of Nature, although the overall outcome of the events may give the appearance of randomness.

So, in my opinion, the continued evolution of the universe in space and time supports the idea that randomness does not exist. To allow its existence, it would be necessary that, occasionally, the laws of Nature changed "randomly" (that's funny), or be suspended for a time, and be reinstated later to the "normal" laws. In other words, for randomness to exist, someone or something must "decide" to suspend or modify the blind and determined behavior of matter for a while. This capricious suspension or modification of the laws of Nature is what could originate randomness; otherwise all material events would be determined from the beginning of the universe.

Religious people have noticed perhaps that what I propose here as randomness is similar to what they consider a miracle, that is, a divine intervention that abolish the laws of Nature for a brief moment. Wounds heal immediately; the dead resurrect; the fish multiply, even dead. However, it is difficult to assume that this suspension of the laws of Nature constitutes randomness embedded autonomously in the behavior of the universe, because the will of God is surely not a matter of chance, but His plan, reasonable and fair. In this way, divine intervention would not be anything other than the completion of some necessary adjustments to the original divine plan to run its course, so it cannot be properly considered as randomness.

Thus, we may think that unless the random suspension of the laws of Nature is a law of Nature itself, randomness does not exist. This may seem paradoxical, and it is. This random suspension of the laws of Nature would be a law of Nature that might say something like: "*Nature, occasionally do not follow your own laws, including this one*". This law, certainly, would prevent us from knowing and anticipating what will happen, but would this be randomness? Can randomness result from a law imposing restrictions on others?

Of course, nobody has observed yet scientifically such a suspension or modification of the laws of Nature. This does not contradict the idea that randomness might not exist. And if chance does not exist, all that has happened and will happen would have been predetermined from the beginning of the universe. Obviously, our supposed individual freedom would also be an illusion, at least with respect to the actions of our material body, for those who continue to think that the human soul exists. If our material bodies follow the laws of Nature, no doubt their actions are determined, unless randomness exists or there is something or someone outside the material dimension controlling matter by changing or suspending the laws governing its behavior.

Finally, a few words about chaos theory might be convenient at this point. Chaotic systems in Physics are systems whose outcome is greatly dependent on initial conditions: small differences in them give rise to very different and unpredictable outcomes. However, despite its unpredictability, the behavior of these systems is deterministic, that is, there are not random factors involved.[5] Therefore, chaotic events reveal our inability to know and predict, not randomness.

It is therefore possible that after the beginning of the universe, everything has been determined, including the wild philosophical speculations of my brain that I clumsily attempted to reflect in the above words. Thus, both life on Earth and the development of our civilization might have been predetermined from the origin of everything. However, even in this case, we continue not knowing whether or not the development of life and civilization elsewhere in our universe is also predetermined.

Returning to this issue, we therefore need to consider that either it is determined or, on the contrary, it is a matter of chance that we are alone or not in the universe. However, since we do not know it, in both cases we could try to evaluate how many civilizations could accompany us. And we could try to calculate it because, although it can be determined from the origin of the universe that I am now writing this, it is almost certain that no one else in the universe is now writing the same words, nor possibly (God forbid!) will someone else write them in the future. In other words, even if a certain event or situation is determined from the origin of the universe, this does not imply, for that reason, that it will happen more often or in additional places. The event could happen only once in the entire universe and only at one point in its existence, like, most likely, the event that is happening right now, dear reader, who are reading these words for the first and, probably, the only time in the history of the universe. You may be the only reader of this book or, on the contrary (may it be God's will!), many will join you. In both cases, this can be predetermined from the origin of the universe, but nobody knows. This analogy leads us to conclude that, even if it was predetermined that there is only one technologically advanced civilization in the universe, or, conversely, that there were many, we may always ask ourselves what is the frequency of other civilizations similar to ours in the universe. Is it great or low?

Similarly, if randomness exists, we also may ask what is the probability (and therefore the frequency) of other civilizations similar to ours in the universe. Their existence might not be predetermined, but their frequency, in statistical terms, could be. We encounter here a similar situation to that found with the atomic decay of radioactive elements. Perhaps we may not know nor predict where we can find a civilization with which to communicate, but we can estimate how many of them may exist in the universe. In conclusion, with randomness or without it, the frequency of civilizations can be estimated by considering the laws of Nature, and the conditions under which the universe evolves, a knowledge that science has sufficiently acquired.

A final consideration before proceeding further is that just as the large number of stars and galaxies making up the universe seems to indicate that the laws of Nature are of such nature (forgive the redundancy) that they determine, or at least favor, their formation and existence, if we find many other civilizations in the universe we should then conclude that the laws of Nature are also of such a nature (forgive again the redundant redundancy) that determine or favor their formation. In other words, we should conclude that the universe is in such a way that not only allows the development of life, intelligence, and consciousness, but that they would be implicit from its beginning. In that case, it might be possible to conclude that the meaning of the universe would be to create intelligent beings who could communicate among them. However, if we do not find more civilizations throughout the universe, or if we determine that the conditions of our existence are so extraordinary that it is extremely unlikely that we are accompanied by other intelligent beings, we must conclude that we are extraordinary, rare, and infrequent beings. In that case our existence may not be due but to a fortunate set of circumstances, not to the fact that the laws of Nature are of such a nature (do not mind the triple redundancy) that they are designed to support the development of intelligence, consciousness and civilization.

The Fermi paradox

The issue whether or not there are other civilizations in the universe has been considered not only by science-fiction writers, but also by first-rate scientists. One of them was no other than the Italian physicist Enrico Fermi (1901-1954), well known for his contributions to developing the first nuclear reactor and to the quantum theory of matter[6]. Fermi was awarded the Nobel Prize for Physics in 1938 for his work on artificial radioactivity and he is considered one of the most important scientists in history.

Fermi is probably less known for his paradox about civilizations in the universe. This paradox, stated in 1950 in a casual conversation during a meal with colleagues, states that if civilizations were abundant in the universe, or at least in our galaxy, they should have contacted with us already. Why then have they not contacted with us yet, not even by the means of robotic space probes, such as those we send to other planets in our solar system and beyond?[7]

Some may think that the paradox is not such because, in reality, extraterrestrial civilizations have discovered us, and even some of their members live among us, as the UFO (Unidentified Flying Objects) phenomenon demonstrates. We mentioned above that science-fiction literature and movies have drawn on this idea numerous times. As examples, it suffices to mention the movie "2001: *A Space Odyssey*" or more recently, "*Avatar*". However, we lack contrasted scientific evidence allowing us to conclude that we are being visited or watched by extraterrestrial civilizations, much less to conclude that extraterrestrials live among us, despite the weird look of some we may cross on the street. In fact, we lack evidence to claim that life exists elsewhere in the universe[8], which obviously includes the existence of another civilization that attempted to communicate with us through radio waves, for example.

However, you may also think, unlike the previous case, that either you have to be a great scientist or completely crazy to articulate such an idea, because the most sensible is to think that extraterrestrial civilizations do not exist or, if they do, they would not travel through space far from their home planet to contact with other civilizations. If this idea had been proposed by a fool, or even by a scientist without a reputation, likely it would have been considered stupid.

However, Fermi knew what he was saying, and most importantly, his colleagues believed that he knew; they had faith in his intellect, and that is maybe why this idea was analyzed in more detail.

First, an estimate was made about how many civilizations could be contemporary to ours in the galaxy. I will not go now into how this estimate was calculated, though we will explain it later (chapter 5) because it is simple. It suffices now to say that experts in this issue (if anyone can be considered an expert in estimating the number of extraterrestrial civilizations), and in particular Paul Horowitz[9] of Harvard University, estimated that there could be a thousand civilizations in our galaxy capable of communicating by radio transmission. They do not seem so many, but in fact, they are a lot.

They are a lot because, assuming each civilization last only a given time, that is, assuming that civilizations, like living beings, are born and die, and they do so at a more or less constant rate, and considering that our galaxy, the Milky Way, is more than thirteen billion years old (as compared to only the four and a half billion years of age for of solar system), and that it is estimated that only one billion years after the Big Bang the galaxy and the universe had generated enough chemical elements to sustain life, then for a thousand civilizations to exist today, about twelve billions civilizations should have existed along the history of the galaxy. How come is then possible that none of them has got here and left no trace, if many civilizations could have arisen in the galaxy millions of years ahead of ours?

Experts believe that there are four possible reasons why, despite existing many civilizations in the galaxy, they have not contacted with us. These four reasons are:

1. Perhaps interstellar traveling is impossible for living beings. In fact, at speeds comparable to that of light, meteorites the size of a grain of sand could destroy any space vessel. In this case, aliens could have never reached us, even if they had intended to. In my opinion, it is likely that even if interstellar travel were impossible for living organisms, it is probably not so for robots, which could have visited us, coming from one civilization or another.

2. Maybe alien civilizations are, in fact, exploring the galaxy, but they have not found us yet. I think, however, that twelve billion civilizations in the galaxy are too many throughout its history for none of them to find us. Of course, some people may think they have done so in the past, and still do so now, but, again, we lack sufficient evidence to conclude this.

3. Perhaps interstellar traveling is feasible but aliens choose not to go out from their planets to explore and colonize outer space. However, I think it is very difficult to consider seriously that the billions of civilizations estimated to have existed in the galaxy have all decided to stay warm and cozy on their planets. In particular, those civilizations that might be in a situation in which their stars are dying would be interested in leaving their planets and attempt to colonize others. This would spur the colonization

of space, leading to the colonization of the galaxy, as we will explain later.

4. Finally, perhaps aliens are already here but have decided not to interfere with us. Again, I think that if civilizations are numerous in the galaxy it is hard to imagine that none of those that may have reached us desire to communicate with us. Since no alien civilization has contacted us, this indicates, therefore, that if aliens have come to Earth, they must not be numerous, that is, they would be individuals only from a few introverted civilizations. If so, this would mean that civilizations are not really as numerous as presumed, in particular because making contact with other civilizations and work with them or try to dominate them has been a constant of human civilizations. I do not believe that civilizations that have devoted the necessary resources to explore the galaxy have decided to go unnoticed if they find another civilization. Why would they go to explore the galaxy in the first place?

Although some of the above reasons could explain why no extraterrestrial civilization has visited us yet, we must also explain why, in the event that extraterrestrial civilizations are numerous but cannot explore outer space or choose not to leave their planets, they appear also not to desire to communicate through radio waves. If an alien civilization discovers that it is impossible, or too expensive, to travel through space, even with robots or other technologies, not for that reason would it be less interesting for that civilization to try answering the important question, both scientific and humanistic (or allienistic), whether or not it is alone in the universe. In this case, this civilization would devote more effort to sending radio signals to outer space, since it would believe that no other civilization could travel to their planet to contact with it, either in a friendly manner or not. In other words, if interstellar travel is impossible, it is to be expected that civilizations would try harder to communicate through radio signals. However, we have not detected radio signals coming from extraterrestrial intelligent beings. The SETI program (Search for ExtraTerrestrial Intelligence) is trying to detect radio wave emissions from extraterrestrial civilizations since 1960[10]. This program has not succeeded, for now, despite the sophistication of the material and human resources devoted to the work, including the distributed computer project SETI@Home[11], which has created, through the shared use of millions of personal computers, the most powerful virtual computer of the world, exclusively dedicated to the analysis of radio signals received from outer space.

Of course, this does not allow us to conclude that intelligent civilizations do not exist, but it does allow us to approach the conclusion that, if they exist, they are not numerous, or they are unwilling to communicate. It is possible that civilizations willingly avoid broadcasting radio waves to indicate their existence to others. If they have a history of wars like ours, perhaps it is the most sensible thing to do, after all. However, again, if civilizations are numerous, it is difficult to imagine that none of them would like to contact with others, not even by radio broadcast. We long for that, why would not others do so too? Moreover, the energy used by the civilizations in its activity dissipates in form of infrared radiation, which can possibly be detected. This has not been the case either, so far. One possible explanation for this may be that civilizations that might exist in the galaxy are as primitive as ours, and have not had time even to reveal their presence by involuntary electromagnetic emissions. However, it is also possible that we are alone. At any rate, some animal was the first out of the sea to crawl along the shore. Perhaps, too, ours is the first galactic civilization and that is the reason why nobody has contacted us: there is nobody out there.

In any case, if we can eliminate each of the potential explanations for the Fermi paradox, we must conclude that we are possibly the most advanced civilization in the galaxy, perhaps the only one to date. Similarly, if we conclude that there are not so many civilizations as experts believe, perhaps that only one, ours, exists in the galaxy, we will have reached, obviously, the same conclusion.

At this point, however, I believe I should offer my apologies to you, because I have hurried too much. I have not explained yet why the Fermi paradox poses a serious problem. I have not explained yet why it is strange that, if alien civilizations are numerous in the galaxy, we have not been contacted by any of them. To do so, please allow me to tell you about the analysis performed by the astronomer Michael H. Hart and the engineer David Viewing in two separate articles published in 1975[12]. The analysis was later extended by the physicist Frank J. Tipler[13] and by the radio astronomer Ronald N. Bracewell[14]. As you can see, this topic has not been addressed by only a privileged mind, but by several of them.

This analysis starts from the assumption, most likely true, as we said, that we have not been contacted by any civilization, so far. It is possible that the UFO phenomenon may be due, as some claim, to alien spacecrafts from outer space, but this explanation is neither the only one, nor the most likely. In the absence of better evidence, we must assume, I insist in this point, that we have not been visited or contacted yet by extraterrestrial civilizations.

And this is a serious problem to keep alive the idea that we are not alone in the universe and that there are many civilizations similar to ours in the galaxy. The reason is this: suppose that a technologically advanced civilization discovers how to travel through space and colonize other planets. Suppose they send a few settlers to the planets that revolve around one or two nearest stars. After the colonies have been established on these planets, the colonists, in turn, send new settlers to the planets around other nearest stars. Likewise, after their settlement, these colonies send again new settlers to the nearest stars...

It is easy to see that if this happened, each civilization would send a wave of settlers from his home planet. How quickly would this wave reach us, starting from any point in the galaxy?

To answer this question, we can be as generous as we want with our assumptions. Let us consider, for example, that planets are colonized at an average distance of ten light-years (the distance traveled by light in ten years, which is about 95 trillion Km., or 59 trillion miles). Let us suppose that spacecrafts travel at an average speed of only 10% the speed of light (our engineers and scientists have already designed, on paper, ships that could travel at twice that speed). Finally, let us suppose that it takes four centuries since the establishment of a colony on a new planet until it sends a second expedition to another planet nearby. With these parameters it's easy to calculate that any civilization could colonize the galaxy in just five million years!, less time than it took humans to evolve from the common ancestor of chimpanzees and our species.

Needless to say, five million years is an amount of time completely ridiculous when compared to the age of the galaxy, which is around ten billion years. Given the differences in the time scales of which we speak, we can even suppose a slower rate of colonization, for instance such that each colonizing stage takes four thousand years, instead of four hundred. That is a time similar to the employed for the development of human civilization from its beginnings, which seems a time too long for each stage of the process we are considering, anyway. However, even in this case, the colonization of the galaxy would happen in fifty million years, instead of five, which is still very little time with respect to the age of the galaxy.

This means that the first civilization with sufficient technological capacity to engage in interstellar travel would have colonized the galaxy in almost an instantaneous time as compared with its age. That is, a single civilization could have colonized the entire galaxy, even before other competing civilizations had time to evolve to the point where they could initiate settlements on their own. If instead of a single civilization we presume the existence of several thousand of them throughout the history of the galaxy, we must conclude that all of it should have been colonized many millions of years ago. This could have happened, moreover, several times throughout the history of the galaxy. However, we have not found any objective evidence about the Earth being colonized in the past, or being visited now.

UFOs and IFOs

It is perhaps worth pausing a little more at this point. First, it is clear that today we have no fluent contact with alien civilizations (often we do not have it either with terrestrial ones). Therefore, if they visit us, they do it in secret. That is what some people believe, people who claim that unidentified flying objects (UFOs) are spaceships from other planets. These ships, should they establish contact with an alleged member of our species, usually do it with some Earth inhabitant of rather primitive mentality, usually in some deserted and lost region. Never a diplomatic alien commission has landed on the gardens of the White House or the Kremlin, or even near Moncloa Palace, Spain, (where the proponent of the idea of a civilization coalition lives), to introduce themselves to the world. So far, this has only happened in science-fiction stories. So, in this case, reality is not stranger than fiction, for the time being. Besides, we will see later why it is unlikely that a first contact between our civilization and aliens happens any time soon.

Of course, some people believe that, in fact, aliens have visited us in the past and, indeed, they were who seeded our planet with life. Well, although I think this is even more unlikely than the possibility of being visited by other civilizations now, even if this were true it would not affect in any way the problem posed by the Fermi paradox. The reason is that the civilization that perhaps seeded life on Earth no longer seems to be among us and, besides, no other civilization has come since then to disturb the evolution of life on our planet. The molecular unit of life, discovered by science, as well as the evolution of living organisms, strongly suggests that the civilization that seeded our planet with life should have visited it when life on Earth began, at least three and a half billion years ago, to leave later without a trace. The existence of such a civilization would be an argument supporting that the galaxy should be teeming with civilizations even today, since at least one arose very early, billions of years ago. However, since then, we have no evidence that other civilizations exist. Why?

On the other hand, is now accepted by scientists that most UFOs become IFOs (Identified Flying Objects). But even for those objects that remain UFOs, there are more plausible and probable explanations than assuming that they are visitors from other planets[15]. It is a common practice of science, by the rationality of their nature, attempting to prove or disprove the most likely hypothesis before embracing the most unlikely scenario. Among the most likely hypothesis for UFOs are, for example, rare or uncommon atmospheric phenomena, which human imagination converts in whatever it desires to see. Among these phenomena we could include spherical lightning or lights caused by earthquakes. Talking about imagination, we cannot rule out the possibility of hallucinations, or various optical illusions, such as the auto kinetic effect, by which a static visual field appears to move[16]. This phenomenon usually happens at night or in low light, when references with which we determine the motion of objects are confusing. Interestingly, the majority of UFO sightings have happened at night. Nor should we forget hoaxes[17] or false reports of sightings. Also UFOs, if they are to be airships, more likely are airships of terrestrial, not extraterrestrial, origin. Some have postulated the possibility of UFOs being experimental aircrafts of the various military powers of the planet[18].

With the above, I do not intend to give an exhaustive account of the possible hypotheses that could explain the UFO phenomenon in addition to that they are alien spacecrafts. I wished, however, illustrating through examples the Occam's razor[19] rule. This rule, stated by William of Ockham (1288-1348), postulates that when multiple explanations are possible, the simpler and more likely should be preferred to others. This seems very reasonable but, in my opinion, it is emotionally unsatisfying, so this rule often is not applied, as it should. Normally, the simplest and most likely explanation is also the least attractive, the least romantic, the least exciting. Many people, weak-spirited, prefer an unlikely emotionally attractive explanation to a reasonably likely explanation. They get personally involved with the explanations, with the ideas; they fall in love, and even marry and sleep with them and, for that reason, they find it extremely painful to abandon them for more reasonable, but less friendly, ideas.

To sum up and not make this point longer than necessary, which, incidentally, I am already doing by mentioning these words, we have no solid evidence to support we are being visited by inhabitants from other planets. In the absence of such evidence, we must conclude, although it is not emotionally exciting, that no one is visiting us. The Fermi paradox is not resolved.

To solve it, if the fact that we have not been visited by other civilizations, despite they might have been very abundant in the galaxy, or even still are, has no plausible explanations, we need then to explore whether it is plausible that we are the only civilization in the galaxy, or that there are only a few civilizations in it, far away and isolated from each other. Obviously, the existence of other civilizations in the galaxy depends on the difficulty with which life might arise on other planets. We will talk about this next.

Notes to chapter 1

1 http://www.britannica.com/EBchecked/topic/184077/electroweak-theory - http://en.wikipedia.org/wiki/Electroweak_interaction

2 http://msc.phys.rug.nl/quantummechanics/ - http://en.wikipedia.org/wiki/Randomness

3 http://www.britannica.com/EBchecked/topic/486231/quantum-mechanics/77521/Hidden-variables - ref=ref894580 - http://en.wikipedia.org/wiki/Radioactive_decay

4 http://en.wikiquote.org/wiki/Niels_Bohr

5 http://www.britannica.com/EBchecked/topic/106013/chaos-theory - http://en.wikipedia.org/wiki/Chaos_theory

6 http://www.britannica.com/EBchecked/topic/204747/Enrico-Fermi http://en.wikipedia.org/wiki/Enrico_Fermi

7 http://www.faughnan.com/setifail.html - http://en.wikipedia.org/wiki/Fermi_paradox

8 http://en.wikipedia.org/wiki/Extraterrestrial_life

9 http://physics.harvard.edu/people/facpages/horowitz.html - http://en.wikipedia.org/wiki/Paul_Horowitz

10 http://www.seti.org/Page.aspx?pid=1366 - http://en.wikipedia.org/wiki/SETI#Where_are_they.3F

11 http://setiathome.ssl.berkeley.edu/ -- http://en.wikipedia.org/wiki/SETI@home

12 Ian Crawford. *Where are they? Scientific American, july 2000, pp 39.* - http://www.sciamdigital.com/index.cfm?fa=Products.ViewIssuePreview&ARTICLEID_CHAR=7CD199C4-9FAF-4059-B2B9-C543C3545CD

13 http://math.tulane.edu/~tipler/ - http://en.wikipedia.org/wiki/Frank_J._Tipler

14 http://www-star.stanford.edu/people/bracewell.html - http://en.wikipedia.org/wiki/Ronald_N._Bracewell

15 http://en.wikipedia.org/wiki/UFO#UFO_hypotheses

16http://www.britannica.com/EBchecked/topic/44824/autokinetic-effect - http://en.wikipedia.org/wiki/Autokinetic_effect

17 http://en.wikipedia.org/wiki/UFO#Famous_hoaxes

18 http://www.rexresearch.com/wingless/wingless.htm - http://en.wikipedia.org/wiki/Military_flying_saucers

19 http://plato.stanford.edu/entries/ockham/ - http://en.wikipedia.org/wiki/Occam%27s_razor

Chapter 2: Lives

Clearly, the number of extraterrestrial civilizations that may exist in the universe, and particularly in our galaxy, where we could more easily detect them, depends on the difficulty with which life can develop in it. This, in turn, depends on the nature of life itself, on what life consists on, and on the conditions necessary for its development. If what it can be consider as life is extensive, if life can arise in various settings and from multiple chemical elements, then perhaps it can arise and evolve almost anywhere in the universe. If, on the contrary, what can be considered as life is more restricted, if its requirements are more stringent, then its development would be less widespread.

Scientists have not yet given a definition of life including all its features[1]. Later I will add my own, but first let us briefly examine the features of life and living organisms, according to the consensus of the scientific community:

1. Organization: Living organisms must have an organized structure, based on one or more cells, the basic units of life, which, in turn, may organize to constitute organs and systems in multicellular organisms.

2. Homeostasis: Living organisms must be able to maintain a constant internal environment; for instance, their temperature and blood electrolyte composition, despite external environment fluctuations.

3. Metabolism: Living organisms generate the energy needed for their functioning through the

transformation of molecules in a network of integrated and controlled chemical reactions. In this transformation, the material parts necessary to sustain them as living organisms are also produced.

4. Growth: Multicellular organisms grow in an organized way, increasing the size of their organs and systems during their development. Furthermore, the species to which an organism belongs to intends to expand and colonize every ecological niche where it can live.

5. Adaptation: Ability to change in response to the environment. This capacity has two components: the ability to adapt each organism possesses during its lifetime (for example, adaptation to seasonal changes, availability of food, etc.), and the adaptation of species to long-term changes produced in the environment, which is crucial in the process of evolution.

6. Reaction to external stimuli: It is actually a more immediate adaptation to rapid changes, sometimes cyclical, occurring in the environment. For example, shutting the eyes to a bright light, or a shift in the position of some plants to better track the sun's motion through the sky.

7. Reproduction: This is the property of life entailing more fun for everyone, and perhaps the property considered as the most fundamental. Reproduction can be of various kinds, such as single cell division (one cell divides into two), budding, or sexual reproduction. In all cases, new cells are generated from preexisting ones.

According to these features of life, I propose the following definition: Life is a *molecular symbiosis* that generates complex (organization) and stable (homeostasis) systems, able to adapt to the environment to maintain their stability (adaptation and reaction to stimuli), to extract from it matter and energy (metabolism and growth), and to self-replicate (reproduction).

Like all definitions, this one attempts to summarize the properties of living organisms in a sentence, which must be necessarily long. In any case, the key concepts included in the definition are those of *molecular symbiosis* and *complex systems, stable over time*, and able to *reproduce*.

As we know, symbiosis is a relationship established between two living organisms so that both benefit from it. Other relationships are also possible in which only one of the organisms draws a benefit, for example in parasitism. In my opinion, symbiotic or parasitic relationships can be established not only among organisms, but also among molecules. For example, the DNA molecule can reproduce only if the necessary enzyme molecules allowing the process of its reproduction are generated from it. These molecules are also reproduced in the process. That is, the entire molecular system is replicated, not just the DNA. All molecules benefit, understanding the benefit as their permanence in time. This is what I mean by the term *molecular symbiosis*. Thus, cells and organisms are not the basis of life, but the molecular systems that make them possible.

We can now ask ourselves about what chemical features the molecules that can interact and collaborate with each other must have to form complex systems (later we will discuss in more detail the concept of complexity proper to life). Can molecular symbiotic systems be formed based on different chemical elements, or can they be formed based on only one element?

Life as we know it is carbon-based. Faced with the problem at hand, which is, let's recall it, whether or not there are other civilizations in the galaxy, we must try to answer the question whether life can be based only on carbon, or whether it may also arise and evolve from the chemical behavior of other elements. Well, despite the many speculations about the possibility that life could be based on the chemistry of other elements, particularly silicon, the answer most scientists currently give to this question is a strong "no". Life based on elements other than carbon is not possible. Why?

Despite what you may have read in many biology and chemistry books, the answer lies not only in the impressive properties of carbon, which we will briefly overview, but also on the properties of the other elements that can interact with it. That is, the answer depends on the molecular systems that can be formed with carbon and a few other particular elements as opposed to the possible molecular systems that could be formed with the rest of the elements. There is much ado about carbon and life, but without other chemical elements, and without their ability to interact with it, life would be impossible. Thus, the important idea for the understanding of the molecular basis of life is to analyze the possible molecular systems that can be formed by the combination of several elements, and the necessary conditions for it, and not just focus on only one. However, it is true that carbon is the central element of life: whereas life could arise perhaps even in the absence of nitrogen, it could never arise in the absence of carbon.

What I just said about the role of the various chemical elements that form the living matter may seem obvious, but it is an idea worth a more detailed analysis. As an analogy to better understand what I mean we can imagine a drawing. Let us imagine a simple design consisting on a black line on a white background. Clearly, the drawing can only exist if we have both, the line and the background on which it is drawn. These two elements, line and background, form a system that only exists if both elements exist together. If only the color black existed, the drawing that I just mentioned could not be drawn because we could not draw any line on a white background in the absence of that color. That is, the possibility of the existence of such a drawing depends on the relationship between two colors. Of course, more complex designs could be drawn with lines of other colors in addition to black. We could now add red, blue, or yellow lines. Each of these colors would add complexity to the drawing.

If, by analogy, we suppose colors in a drawing are as atoms in a molecule, starting from a given atom we may add others to add complexity to the molecules that could be formed. But whereas you can add colors to the drawing following no rule at all, this is not the case with atoms to form molecules. Returning to the drawing analogy, the existence of rules would imply, for instance, that a black line could not cross other lines of certain colors, and similarly, that a yellow line (a different atom), likewise could not cross lines of the same or different colors as those the black line can cross. That is, each color would have to follow particular rules that would determine which other colors it can cross in the drawing.

With this analogy we can perhaps understand why carbon is so important for life. Let's suppose we have a yellow line whose associated rule is that it can only cross blue lines, and nothing else. It is clear that multi-colored drawings based on yellow lines are impossible. All the yellow lines present in the potential drawings could be crossed only with blue lines, which would severely limit the ability to make different designs, or at least designs of varied colors.

Let us imagine, however, a black line, whose associated rule is that it can be crossed with other black lines and also with white, blue, green, and red lines. Clearly, the possibility of elaborating different designs of different colors is far greater. In other words, while the first situation only allows the making of simple drawings, the second allows the elaboration of much more complex drawings. This condition is essential for living systems, which, by definition, are complex, even the simplest of them (the simplest virus has a high level of complexity).

Let us focus now on the atoms, in particular on carbon. This element is similar to the black line mentioned above, in the sense that it can be crossed with many other elements, beside itself, in particular with hydrogen, nitrogen and oxygen, but also with sulfur, chlorine or fluorine, among others. However, the first three elements together with carbon are the most abundantly found in the molecules of living organisms. With carbon and the other three elements, billions of different molecules can be formed. This is also because each carbon atom can bind to four other atoms at the same time and, in particular, it can bind to itself to form long chains, linear, branched or cyclic. Returning to the analogy of the drawing again, with four colors and their combination rules billions of different patterns could be drawn, because, in addition, the lines are not restricted in their lengths.

Thus, the ability of carbon to link to itself and to other elements, particularly to hydrogen, nitrogen and oxygen, is essential to allow the generation of diverse and complex molecules. But apart from this property, carbon has another very important feature: whereas it is able to bind these atoms easily, it can also separate relatively easily from them.

Why is this property important for life? It is important for several reasons. First, if carbon bound chemically to, for example, oxygen so strongly and easily that it could not separate from it, all the carbon on Earth would then be bound to oxygen. All carbon would be in the form of CO_2. Besides the huge greenhouse effect our planet would suffer, there would be only a simple molecule, CO_2, so stable, that is, so difficult to break, that more complex molecules could not form. Fortunately, even though CO_2 is a stable molecule, and oxygen binds to carbon tightly, it does not bind to it so strongly that prevents its separation from carbon, thus allowing it to bind to other elements, such as nitrogen, or hydrogen, to which carbon binds even more strongly than to oxygen. Plants carry out the separation between carbon and oxygen during photosynthesis, a process in which CO_2 and water (H_2O) are converted into carbohydrates (empirical formula CH_2O), so necessary to sweeten life.

Another reason why it is important that carbon does not bind very strongly to one particular element in relation to others is that the energy needed to break out the bond between, for example, carbon and oxygen, is similar to that released in the formation of, for instance, the bond between carbon and nitrogen, or carbon and hydrogen. Being of similar magnitude, these energies can be used to the rupture and formation of bonds between carbon and other elements so that the final energy balance is not very high. This enables the molecules constituted by carbon, hydrogen, nitrogen and oxygen, i.e. all the molecules of life, to be easily transformed into each other, thus forming a complex molecular system in which the exchange of matter and energy is relatively easy. This property is absolutely essential to carry out the chemical reactions that generate energy and produce the components of living matter from other organic molecules, or from CO_2 and water.

Thus, the easy way in which carbon forms and breaks out bonds with other elements is essential to form the diversity of molecules proper of life and to the generation of greater complexity. Before proceeding further with other topics, it is convenient to analyze the idea of complexity in more detail. The Merrian-Webster dictionary defines "complexity" as *"quality or state of being complex."* In addition, it defines "complex" as *"a whole made up of complicated or interrelated parts"*. Well, in my humble opinion, this is not a good definition for the concept of complexity. For an entity to be complex it is certainly necessary that it be made up from diverse elements, but this is not enough. To understand why, suppose an object consisting of five parts and compare it with another made up of ten parts. In principle, the object consisting of five parts would be simpler than that formed by ten. But suppose that the five parts making up the first object can be joined in various combinations, while this cannot happen with the parts making up the second object, which can only bind to each other in a particular way. Obviously, this would mean that the first object would be just one of a number of possible objects that could be formed by joining the five pieces in different ways. The second object would be, on the contrary, the only one that can be formed by joining the ten pieces. It is, therefore, in the sense of comparing the actual existence of an entity against the possible existence of other entities formed by various combinations of the same components making it up how we have to understand complexity. Thus, not only the molecules of life are complex because they are formed by a variety of chemical elements, but because these elements can be linked together in a huge number of different ways.

With these ideas in mind maybe we can now better understand why life is based on the chemistry of carbon. Carbon is the chemical element with the appropriate properties, in relation to itself and others, which enables it to form complex molecules that can interconvert with relative ease, allowing the generation of complex symbiotic molecular systems, which are the basis for life processes. Are there other elements with similar properties to carbon? Is it possible that complex molecular systems in the universe could be formed from other atoms? In particular, would silicon, also capable of binding up to four other atoms, be able to build similar systems?

On paper, it would be possible, perhaps, but in the reality of our universe, it is impossible or, at least, very unlikely. The reasons are several, and we will try to explain them all (all that I have been able to unveil) because this way, I believe, we will better understand why life in the universe can only be based on carbon, and why if we find another civilization in the future, the organisms belonging to it will also be based on this chemical element.

Silicates

The first reason is that silicon binds to oxygen much more strongly than it binds to other elements, in particular to hydrogen or nitrogen. This is why almost all the silicon in the Earth's crust is in the form of silica (chert, quartz, sand...) or silicates. Silica and silicates are the most abundant minerals of the Earth crust and mantle[2]. The silicon atom being surrounded by four oxygen atoms bound to it by very strong bonds characterizes the molecular structure of these minerals. The bonds are so strong, and silica so stable, that the silicon atom "prefers" binding to oxygen rather than to any other element, including itself. This means that, in the presence of oxygen, long molecular chains of interconnecting silicon atoms cannot form, unlike those formed by carbon. The impossibility to form long chains of silicon atoms in the presence of oxygen makes the formation of complex molecules based on silicon impossible. Only simple molecules, mainly consisting on silicon and oxygen, will form.

Well, you would think: so what? All we have to do to make silicon binding to itself and to other atoms to form complex molecules is to get an oxygen-free environment. Surely, the universe, being so vast, contains planets here and there with no oxygen, in which silicon could form chains similar to those of carbon, which will enable the generation of complex molecular systems. It is possible, therefore, that in those planets very different life forms will develop, and (why not?) they may also evolve to form a technological civilization.

Well, it's not so simple. It is true that our universe is vast and wide, and we can imagine a huge variety of possible planetary conditions, since, surely on this we will agree, is on planets where life can develop (at least, life worth trying to contact with). However, not everything is possible, even in the vast universe, and it is very difficult, if not impossible, to find silicon in an oxygen-free environment. Why?

The reason is that not all chemical elements are equally abundant in the universe. If it were so we would have similar amounts of gold, platinum or uranium on our planet. Fortunately or unfortunately, this does not happen. It turns out that oxygen is nothing less than the third most abundant element in the universe. After hydrogen and helium, the only two primordial elements formed in the Big Bang, oxygen is the most abundant. However, silicon is only the eighth most abundant element in the universe. Not bad, but five places lower in the ranking of universal abundance. In fact, for every atom of silicon, there are more than nine oxygen atoms in the universe[3].

This means that there is enough oxygen in the universe to bind to all the silicon atoms, and enough of it left over still to bind to other elements, as, for example, to hydrogen, to form water; or to iron, to form oxides of this metal. This is indeed what happens on our planet, in which, oxygen is bound to almost all the silicon, despite this element being the second in abundance in the Earth's crust after, precisely, oxygen. The few silicon atoms that may remain unbound to oxygen lie in the depths of the Earth, where it is very difficult that it may intervene in the complex chemical reactions necessary to form living organisms. Additional reasons, which we will explore later, increase the difficulty of this possibility. That is, it is twice extremely difficult that silicon binds to itself, forming complex molecular chains.

Still, you may argue again: so what? Surely among the millions and millions of planets that appear to exist in our galaxy, according to recent research, we will find one in which oxygen is absent, despite its abundance, or at least in which it is less abundant than silicon. Whereas I will not deny that this may be true, I deny that it is likely. That is, if it happens it does not happen often and there will be few, very few planets I dare to say, where silicon dominates in relation to oxygen. Moreover, not all of these will have the conditions of temperature, pressure; etc.., so that a silicon-based life could develop.

The reason why very few planets will have more silicon than oxygen is that silicon and oxygen are not formed at separate points, but together in the same regions of the universe. Once formed, they are released to space at the same time, so they come out mixed. Because oxygen is formed much more abundantly than silicon, this mixture, when it gathers to form the planetesimals that later will become true planets, will always be richer in oxygen than in silicon. Silicon, inevitably, will end up bound to oxygen, because, as already mentioned, it has a great affinity and binds very strongly to it.

But where in the universe do the chemical elements that are now in the Earth and in our bodies form? The answer may surprise many: at the core of stars. Yes, it is in the center of stars where the conditions of pressure and temperature are those required to ensure that the primordial elements, hydrogen and helium, fuse together and generate the nuclei of new elements, including carbon, oxygen, and nitrogen, which together with hydrogen are the basis of life as we know it.

Obviously, silicon and oxygen, and all other elements found on Earth, had to leave the core of the star where they were formed. Releasing out the chemical elements formed at their cores is not an easy task for the stars. In fact, most of them are unable to do it. Special stars can only perform this task, large stars which, because of their large masses, rapidly fuse hydrogen and helium and form many other chemical elements. Due to the speed of the undergoing nuclear fusion at their cores, these stars live a short life as compared to others, and soon arrive to their deaths, which occur in large explosions that eject most of their matter into space. When they die, these massive stars are called supernovae[4], because they appear in the sky as new stars whose brightness can be even greater than that of the galaxy to which they belong. However it is not the briefness of their lives and their spectacular death what is important. What is important is that their short lives have been creative: they have created new elements and, after their deaths, they leave a vital legacy to their surrounding areas, which makes possible the birth of

63

new stars. Around them, planets may form on which life might develop.

Thus, the chemical elements that are now inside your body, inside the regular or electronic book you hold in your hands and the chair or sofa that is holding your buttocks, were formed inside a star that exploded. Clearly, the brutal explosion had the sensibility not to separate the elements formed in the star and, on the contrary, produced a large cloud of gas and dust that flew out of its core at high speed. From this cloud, from this nebula, a new star arose, our Sun, and also the planets revolving around it. One of them is Earth, where we are. Wonderful, isn't it?

During the formation of the Sun and planets, when the rate of expansion of the nebula decreased due to the effects of gravity, then some zone by zone partition of the elements occurred. Some areas of the nebula contained more mass than others, and therefore their gravitational pull was greater. These areas recollected more hydrogen, the lightest element, which requires a greater gravitational force to be retained. In these areas the Sun and giant planets, like Jupiter and Saturn, formed. Other parts of the nebula, with less matter density, could not hold as much hydrogen, and only retained heavier elements, like oxygen and, fortunately, carbon. This applies to our planet, which is able to retain hydrogen only because it binds to oxygen forming water; earth's gravity is not sufficient to retain it as a gas. However, despite these gravitational effects, segregation of the elements was very limited and most of them remained largely mixed.

So, returning to the issue of oxygen and silicon atoms that may be present in planets formed in our galaxy and the entire universe, these chemical elements have formed as the rest of them, in the core of stars which died as supernovas. However, starting from the hydrogen and helium, the only two elements that, as stated above, were formed in the initial Big Bang, not all elements are formed in similar amounts. Some form more easily than others. This is the case of oxygen, which, as we said, is now the third most abundant element in the universe. This means that it has formed more abundantly than silicon, than carbon (the fourth most abundant element in the universe) and than nitrogen, which occupies the seventh position in the ranking of abundance. Of course, hydrogen is still, by far above the rest, the most abundant element in the universe, followed by helium.

On the surface of Earth there are almost twice as many oxygen atoms as those of silicon, carbon, hydrogen and nitrogen together[5]. This means that if oxygen strongly bound all these elements, all of them would be in an oxidized state, such as SiO_2, CO_2, H_2O and NO_2, as there is more than sufficient oxygen to oxidize them all. However, this only happens in the case of silicon. Not all hydrogen, carbon and nitrogen atoms are bound to oxygen, and this is, as we have said, because these elements bind to each other with similar strength. Not so silicon. Silicon is faithful to oxygen. Once bound, it does not part from it (at least at the temperature conditions found on the surface of the Earth), so it is completely oxidized.

Thus, all discussed above explains why, despite the fact that silicon has similar properties to carbon, silicon may not be the fundamental atom for life processes based on molecules formed mainly by it and other elements. In the presence of the much more abundant oxygen, silicon is "sequestered" by it and cannot intervene in other chemical processes, much less in chemical processes of the complexity of those found in life. However, carbon, smaller and more promiscuous, cannot be hijacked by oxygen, and it can form different molecules that, in fact, have created life.

However, despite this, in all chemical fairness, there is a possible way that may liberate silicon from oxygen. This possibility is offered by hydrogen. Oxygen binds to hydrogen a little more strongly than to silicon, thus forming water. This would mean that if a planet retained sufficient amounts of hydrogen, all oxygen could be preferentially in the form of water, rather than forming silicates. This would free silicon to form long chains, such as those formed by carbon. Better yet, the affinity with which, in the absence of oxygen, silicon would bind to carbon, nitrogen, or hydrogen, the most abundant elements to which it could bind, is similar. This means that, as in the case of carbon, bonds of silicon with these elements could be formed and destroyed with relative ease, thus enabling the transformation of molecules needed for the molecular symbiosis proper of life. Thus, we could perhaps imagine life based on silicon, provided that hydrogen, more abundant than oxygen, is sequestering this element, letting silicon free to combine with other elements.

This raises, however, a serious problem, even if this hypothetical planet were located at the optimum distance from its star so that life may develop. The problem is that the molecules formed by chains of silicon and hydrogen, similar to hydrocarbons, decompose rapidly in water, because silicon can form four very strong bonds with oxygen, but oxygen forms only two with hydrogen, so the energy balance remains favorable to the binding of silicon with oxygen, even if the later has been previously sequestered by hydrogen. The hydrogen of this hypothetical planet could have bound all the oxygen to form water, but this would mean that water would be abundant. Under these conditions, life based on a silicon chemistry could be formed only in places on the planet free from water, and of course, free from oxygen, which probably would prevent living organisms reaching the surface, since water would likely accumulate on it, either in large oceans or in the form of water vapor in its atmosphere. Under these conditions, it would be even harder to imagine how these organisms might evolve to create a technologically advanced civilization.

This means that whereas water is needed to allow carbon-based life, it should be absent in the case of life based on silicon. This is unlikely, given the abundance of oxygen and hydrogen in the universe, the first and third element in order of abundance, respectively, as stated above.

Entropy

Water leads us downstream to discuss another absolute requirement for the development of life. This requirement is that life needs a molecular medium that can increase its disorder, thus allowing the arrangement in complex systems of the molecules making life possible. That is, besides the formation of complex molecules, of molecules composed of many different elements, you need a medium in which these molecules can collaborate with each other and get organized. Without that environment, life is impossible.

The fact is that there is a major difficulty for the development of organized complex molecular systems and, therefore, for the development of life. This difficulty has been discussed multiple times and, as many other science topics difficult to understand, it has even been used to argue for a supernatural intervention to explain the existence of life. This difficulty is called the second law of thermodynamics.

This general law, evidenced in all systems studied, also in living systems, derives from a universal behavior of physical and chemical systems. This behavior is such that along the temporal evolution of these systems, if they are not in a state of equilibrium in which evolution has stopped, the disorder of the universe increases. In more technical terms we say that the quantity called *entropy* of the universe increases.

For instance, if in two compartments separated by a gate we introduce nitrogen into one and oxygen into the other, the initial state is well ordered. If we then open the gate separating the compartments, the two gases may freely pass from one compartment to another, and that is what will happen. The final state will be a homogeneous mixture of the two gases, a more disorderly state than the initial one. This is what happens in Nature, whereas it could perhaps happen instead that the gases stayed in each compartment, unmixed. Furthermore, the opposite phenomenon, even if possible, it has never been observed. That is, it has never been observed that a mixture of two gases get ordered by the spontaneous separation of the molecules of either gas in different areas of the vessel containing them. Thus, the physical and chemical systems tend to increase disorder, often irreversibly; i.e., they tend to increase the entropy of the universe[6].

This is very important, because spontaneous processes occurring in Nature are those in which entropy increases[7]. This is the reason why some have argued that life cannot be a spontaneous process in the universe, since it means order and organization. The decrease in entropy, and therefore in disorder, that living beings display has been used as an argument for divine intervention and vitalism[8]. However, science now fully understands that life is indeed a spontaneous process, since life increases, it never decreases, the entropy of the universe. Here's how.

It turns out that the fact that chemical and physical systems tend to increase the universal disorder does not mean that, in some cases, the order of atoms or molecules making up a material system cannot increase. Ice offers us a familiar example of increasing order. When cooling down water below zero degrees, it freezes to form ice, and when heated, ice melts. This process is reversible and we can observe it every time we let melt in a gin and tonic the ice cubes we have prepared in the freezer. Surely, you know what I mean. But do not run now to the fridge to prepare yourself a gin and tonic. It is not necessary to drink alcohol to understand what follows. Please continue reading, which is also fun and enjoyable.

Water molecules in the liquid state are much more disordered than in the solid state. They can move everywhere within the fluid, colliding with each other, binding to other water molecules momentarily, and separating from them the next moment. They are free, in short, to move throughout the entire space occupied by the liquid, be it a drop of water, or the ocean.

However, into the ice water molecules are not free. They are all connected to each other forming a three-dimensional network. Each water molecule occupies a certain place and cannot move. Granted, ice is a much more orderly structure than liquid water. But if all systems tend towards disorder, how it is possible that ice may exist? In other words, how once ice has melted, and its molecules have disordered, these same water molecules can re-organize into ice again, if disorder always increases?

The solution to this apparent paradox is that if disorder must increase, it must not necessarily do so exclusively inside a particular system, but in the entire universe. This means that you can sort out the atoms or molecules of a particular chemical or physical system if, in doing so, *the disorder of atoms and molecules outside this system increases in a larger amount* (although, sometimes, an equal amount suffices). *An exchange of order and disorder among some parts of the universe may therefore happen.* For this reason, some parts of the universe can be ordered at the expense of a greater disorder of others.

When we introduce, full of water, the ice cube mold in the freezer, and the water freezes, the disorder of the water molecules decreases, but the disorder of the air molecules and those of the walls of the freezer around it increases, although we do not see it. How is this exchange of order by disorder produced?

Let's see. To freeze the water we must cool it down, and this implies extracting kinetic energy, i.e. the energy of motion, from its molecules. When water is liquid, at room temperature for example, its molecules are moving at high speed, which prevents them from binding to one another for long periods, so they cannot be set in order. To be able to bind to each other, be set in order, and form ice, the speed of the water molecules must decrease.

This decrease in motion, or kinetic energy, can only be achieved if the water molecules transfer their energy to other lower energy molecules, or if they release energy as some kind of electromagnetic radiation, such as infrared light. Energy cannot disappear, it is always conserved (this is the first law of thermodynamics), and the only way in which a system may decrease it is that another one increases it. In the case of freezing water, the kinetic energy of its molecules is gradually transferred to surrounding air molecules and the rest of the freezer.

The transmission of this energy occurs by collisions between molecules of the surface of the water and air molecules, or the molecules of the wall of the water container. The water molecules introduced in the freezer are warmer than those of the air around them. This simply means that they have more kinetic energy than the air molecules and move faster than them. Water molecules from the surface, when colliding with colder (i.e. slower) air molecules, transfer to them part of their energy, and get cooler. Many millions of molecular collisions occur every second. In each one of these, water molecules lose energy and get cooler, and air molecules gain energy and get hotter. The water molecules inside the liquid, now hotter than the surface's, also collide with them and in turn lose energy, cooling down, but warming up the surface's molecules again, which will again collide with air molecules in the freezer, cooler than them, thus cooling down in turn. This state of affairs is repeated until all the water molecules have lost enough energy, so that freezing starts. A similar process happens when water molecules collision with the molecules of the wall of the water container, which also transfer the kinetic energy to the air and walls of the freezer in a chain of molecular collisions.

Freezing can occur because the less energy the water molecules possess, the greater the likelihood they get arranged, since they move more slowly and thus they may bind more easily to neighboring molecules (water molecules can bind to each other by electrostatic interactions). When their kinetic energy is low enough, all the molecules bind to each other and freeze in one position, forming ice. On the contrary, air molecules, having gained kinetic energy from the water, get more disordered. We will not mention here the increase in entropy generated by the entire mechanism of the freezer motor, required to achieve the low temperatures inside.

Thus, the increase in order of the water molecules when they freeze into ice is achieved by increasing the disorder of other molecules. Exactly the same phenomenon occurs in the formation of ordered systems proper of life, such as cell membranes or other cellular organelles.

However, in the case of life we have an additional problem. This problem is easy to understand if we consider the conditions necessary to keep water frozen. If we remove the ice cubes from the freezer to add them to the gin and tonic, they will melt and get back to the liquid state, losing the order acquired. To prevent this, it is necessary to keep the water at a temperature below the freezing point. That is, to maintain an ordered state is necessary to maintain the conditions that make it possible. If these conditions are absent, order is lost.

This means that to maintain order and organization in living systems, it is required that certain conditions remain constant. It is not enough to sort molecules out here and there to form ordered structures if they cannot be sustained. Thus, conditions are required so that once order and organization are acquired, they do not get lost.

Which molecules have the ability to constantly scramble to transfer order to the molecules forming part of living systems? The answer is simple, and you know it: water molecules. There is much ado about the properties of water as a liquid, as universal solvent, as the major component of living organisms in which the metabolic reactions are carried out, and more wonders of the underwater world, but little is ever talked about the properties of water as an *absorber of disorder*, as the substance that makes possible the maintenance of order in living systems. It's time to do it justice also for this reason.

To understand what, in my opinion, is the most important property of water, we must analyze a little longer the chemical nature of liquid water and explore how this wonderful liquid works, explain what it does to dissolve so many substances, and, especially, what happens to water molecules when they encounter substances that cannot dissolve. Although we speak of water, what follows may seem somewhat dry, but I hope that, with the help of a gin and tonic if you need it, you may come along with me anyway.

As surely you know, even if you only drink wine or beer, the water molecule consists of one oxygen atom bound to two hydrogen atoms. What you are probably less familiar with is that the hydrogen atoms are placed at the vertices of a slightly irregular tetrahedron, with the oxygen atom at its center. A tetrahedron is a pyramid whose base and sides are equilateral triangles, and so it therefore possesses four equivalent vertices. Hydrogen atoms occupy two of these vertices, and the other two are empty. Well, not totally empty: four electrons from the oxygen atom are located on them, two at each vertex[9].

This molecular structure causes an imbalance of electrical charges. At the two corners where the hydrogen atoms reside, since atoms have positively charged protons at their nuclei, there is an excess of positive charge. However, at the other two vertices of the tetrahedron, where the electrons are placed, there is an excess of negative charge. That is, half the water molecule is positively charged, the other half is negatively charged.

Even if you you're halfway through your gin and tonic, you surely remember that electric charges of the same polarity repel, whereas those of opposite polarity attract. Because each water molecule has a positive and a negative part, these two parts from two different water molecules will attract each other.

The attraction that water molecules exert to each other is crucial to understanding the properties of this substance as a liquid. If water molecules did not attract each other, water would be a gas at room temperature. This is what happens with oxygen, nitrogen, or carbon dioxide, formed by two or three atoms heavier than hydrogen, despite which these substances are gases at room temperature. Not so water, which, due to the affinity among its molecules, form molecular aggregates with sufficient mass as to remain liquid at temperatures at which water should be in a gaseous state.

The distribution of electric charges that water molecules possess is also fundamental to confer it its properties as universal solvent. Because water molecules can establish electrostatic interactions with other electrically charged atoms, water can dissolve the so-called polar substances, i.e. those possessing electrical polarity, as it happens with the water molecule itself. For the same reason, water can dissolve many salts, since they all consist of charged atoms.

But the most important factor derived from all this is that the electrostatic interactions that water molecules can establish among themselves are a great source of order, which can be transferred to some other molecules immersed in water. To understand this better, let's examine what happens in pure liquid water. In this environment, the water molecules interact electrostatically with each other. They form temporary unions between the vertex of a water molecule occupied by a hydrogen atom, which has an excess of positive charge, with the vertex of another water molecule occupied by electrons from the oxygen atom, which has an excess of negative charge. These interactions, called *hydrogen bonds*, are not limited to two molecules, and many water molecules can interact, using the four corners of the tetrahedron where the electrical charges are located. Thus, molecular chains and clusters are formed with thousands or millions of water molecules. But these clusters are not static. Once formed, they quickly disperse to form others. Water molecules in the liquid state, although they interact with each other, are free to break this interaction and to form other interactions with different water molecules.

This freedom of the water molecules to move everywhere, joining others and splitting out next moment, occurs in all parts of the liquid, except at its surface. Water molecules located there cannot interact with air molecules. If they could do so, they would establish interactions with them, and would not need to be ordered at the surface. But they cannot, because air molecules are neither charged nor polar in nature. This limitation causes that at the water surface the molecules form a two-dimensional network, with molecules linked to one another by hydrogen bonds, which in this case do not break so easily, because they only do so if after the break other hydrogen bonds can form. At the liquid surface, this is more difficult than inside. For this reason, the molecular network at the water surface is quite stable and strong. It is able to hold the weight of insects as large as water striders, which do not swim or float in water, but literally walk on its surface.

This molecular network occurring on the surface of water entails a limit to the possible disorder of the molecules located there. On the surface, the water molecules must be ordered, because the intermolecular interactions among them dominate over their potential disorder, i.e. they are more important, more powerful, that the tendency to disorder. This feature of water molecules, able to interact only with polar substances, and therefore being disordered and free in the presence of these molecules, also means that water molecules are not so free if they are faced with molecules with which they cannot interact, as it happens in the case of air molecules at the surface.

This is also the case with organic molecules, i.e. those formed by chains of carbon and hydrogen, such as hydrocarbons and fats. These molecules are not polar, i.e. do not have a skewed distribution of electric charges and therefore cannot interact with water, unless they have a polar part, i.e. with electrical charge. However, even if they have it, their non-polar part still will be unable to interact with water.

The impossibility of interaction of water molecules with non-polar organic molecules creates a situation similar to that found at the water surface in contact with air. Recall that in this case, water molecules were arranged to form a two-dimensional molecular network at the surface. The same type of molecular arrangement occurs around an organic molecule that is introduced into water. In fact, an easy way to consider what happens when organic molecules are introduced into water is to assume that very tiny air bubbles are introduced instead. Unable to interact, the water molecules get arranged around them, forming a layer of ordered water molecules surrounding the organic molecule, as they would surround an air bubble.

This is bad news to the entropy of water, as it is forced to decrease. For example, the mere addition of a drop of oil to a glass of water results in an ordering of water molecules around the surface of the oil drop. The number of well-ordered water molecules depends on the amount of surface offered to water from the oil drop, which in turn depends on the amount of oil added. In any case, however, the number of ordered molecules is the lowest possible, since the disorder tends always to be the highest.

You may ask now: what does all this have to do with the order of living systems and with water being an absorber of disorder? To understand this, we must analyze what happens when not one, but two identical drops of oil are introduced into water. Obviously, water molecules are then forced to be ordered around the two drops of oil, and the number of molecules ordered will depend on the amount of surface from both drops in contact with water. There is a geometrical law that tells us that *the surface of objects grows according to the square of its average radius length, while the volume does it according to the cube of that amount.* This means that surface grows more slowly than volume, as we can confirm by inflating a balloon and measuring the increase of its surface in relation to the volume of air introduced. Therefore, a spherical drop of oil double in volume in relation to another does not have twice the surface, but a little less. This is very important for understanding the behavior of water in contact with organic molecules with which it cannot interact. Following its "desire" to get disordered, the oil-water system tends to the maximum entropy, i.e., to achieve the highest number of disordered molecules. This number will be achieved when the interface between oil and water reaches a minimum, and this happens when the two drops of oil unite into a single drop, which eventually ends up happening. Thus, although the new drop will have twice the volume it won't have twice the surface. The total surface facing water will be smaller than before, which means that a number of water molecules will have been freed from the obligation to be ordered around the surface of the two drops of oil. The number of water molecules in a disordered state will have increased and, with it, the entropy as well. In exchange, the oil molecules are more ordered now: before they were separated into two drops but now they are together into only one. Once together, they will not separate. As it never has been observed that a gas mixture separates spontaneously, it never has been seen that a drop of oil, or organic material, in water separates spontaneously into two drops. The reason is the same: the second law of thermodynamics tells us that spontaneous processes are those in which the disorder of the universe increases, or at least it does not decline.

We can now better understand why two liquids that cannot be mixed, such as those contained sometimes in hair-restorer bottles (with two phases, organic and hydrophobic the first and aqueous the second, that we need to shake and mix before applying to our hair), spontaneously separate in the two phases again after shaking and let them rest for a while. It might seem that this process, happening spontaneously, is a violation of the second law of thermodynamics, because what we've mixed by shaking, and therefore seems more disorderly than before, appears to organize spontaneously when we let it rest. However, this is just apparently so, because what happens when the organic phase and the aqueous phase separate is that water molecules get more disordered, not more organized.

The reason is that when we shake the bottle of hair-restorer and mix up the two liquid phases, we cause the formation of millions of droplets of organic liquid that are dispersed into water. These droplets of organic liquid phase are now forcing the water molecules to organize around them. The energy used to shake the bottle is, therefore, used to order the water molecules around non-aqueous molecules. It is sound, because energy is always required to order something, as anybody with a husband knows. When, after shaking the bottle, we let it rest, the non-aqueous droplets coalesce, as explained above, resulting in greater disorder of the water molecules that were arranged around them. Thus, the separation into two phases of an aqueous and organic mixture happens according to the second law of thermodynamics, not in contradiction to it.

The same situation occurs, even more easily, if we consider isolated organic molecules. Around them, an orderly envelope of water molecules is established. But this wrapping does not prevent the organic molecules from being pushed here and there within the liquid. Ultimately, it is certain that two organic molecules, each with its covering of water molecules, will encounter each other. At that time, the water sheaths break up to form a larger envelope of water molecules, encompassing the two organic molecules. In this process, water molecules are released that can be more disordered than before, so entropy increases.

If we add more and more organic molecules to water, they will be finally collected in an orderly fashion, forming diverse molecular structures (depending on their precise chemical nature). One of them is the lipid bilayer, which forms the membranes of all cells and enables the cell's organization in different compartments and internal organelles, as well as the functioning of the nervous system (and, therefore, intelligence), which requires an electrical potential between the two sides of the neuron's lipid membrane for the communication among them.

Thus, when submerged into water, organic molecules are arranged in complex structures because, in the process, water molecules are disordered in a greater amount, and total entropy increases. The amount of entropy that increases due to the disorder of water molecules is greater than the amount of entropy that decreases due to the organization of organic molecules, which is why ordering of organic molecules amid water occurs spontaneously. This ability of water is essential to maintain the order of living structures, not only in regard to biological membranes, already mentioned, but also in regard to the proper folding of proteins and the double helix structure of DNA, perhaps the most famous molecule of life.

Therefore, in a universe such as ours, which tends to the maximum disorder, the chemical nature of water and organic molecules is necessary so that spontaneously ordered complex systems, typical of life, can form. In short, for life to exist, water is as important as non-water, this later understood as the molecules that cannot interact with water, and are ordered at the expense of the increase disorder of water molecules. As for a drawing for exist there must always be a figure and a background, paper and non-paper (ink or paint), for life to exist there must be a figure and a background: the background is water; the figure, the molecules that evolve in it.

In addition, it is also essential that the non-water molecules be also in the liquid state. If the non-water molecules were solid at the same temperature as the water is liquid, they would precipitate, and ordered structures, in particular biological cell membranes, whose fluidity is crucial to enabling the exchange of matter inside and outside the cell, could not form. This implies that for life to develop in the aqueous environment, the molecules that can be ordered within it must have physical and chemical features allowing them to be in the liquid state. This requirement limits the type of water insoluble molecules that can be ordered in it to allow the emergence of living systems. These molecules appear to be only those formed by carbon chains, combined with some other elements, mainly oxygen, nitrogen and hydrogen.

After this rather long digression we can now return to analyze whether life based on silicon molecular structures could exist. We can say now that even if a planet rich in hydrogen, in which all oxygen were sequestered by it, existed, thus theoretically allowing the formation of silicon-based complex molecules, this would not suffice for the emergence of life. It would be also required that these molecules be in a medium absorbing entropy. That medium could not be water, since the oxygen in it would break silane molecules (chains of silicon). In addition, it would also be necessary that the physical state of the molecules of life, and the medium in which they were, was liquid, the only physical state able to absorb molecular disorder quickly, and the only one allowing the development of chemical reactions at a high enough rate as to allow them to form part of living processes. For what it is known about silicon chemistry and the stability of the molecules that could be formed with it, it is impossible that molecules as complex as proteins or DNA could be formed and maintained in a liquid state, dissolved in unlikely liquids, such as ammonia or methane, or even hydrogen sulfide, H_2S, because they could not be dissolved in water for the reasons stated above. Another serious problem is how the energy exchanges specific to the chemistry of life could be managed with silicon-based molecules. The oxidation of silicon compounds also releases energy, even more than that produced in the oxidation of carbon compounds, but remember that oxygen must be absent, or sequestered in the form of water, to allow the formation and stability of complex molecules based on chains silicon.

At this point, one may wonder whether carbon-based life would be possible in other liquids used as solvents and absorbents of disorder. The only substance, which, in my opinion, is worth considering, is ammonia[10]. Ammonia is formed by a nitrogen atom bound to three hydrogen atoms. The ammonia molecules are structurally similar to those of water, and can also form hydrogen bonds among them. For this reason, ammonia could also serve as an absorber of disorder. However, the hydrogen bonds formed among ammonia molecules are weaker than those formed with water molecules, so that ammonia requires lower temperatures to remain liquid. Ammonia remains liquid between temperatures of –77°C and –33°C (–106.6°F and –27.4°F). At these low temperatures, organic molecules complex enough to be part of living systems would be in a solid state. More so would even happen with molecules composed by chains of silicon, less volatile than those based on carbon. Therefore, if life molecules must be carbon-based, life can only develop in an aqueous medium.

In summary, the difficulties of a life based on carbon and evolving in aqueous media are great enough to allow considering that life based on silicon, or life developing in another solvent and absorbent of disorder different from water, would be likely. If possible, it seems clear that those kinds of lives would imply even more improbable scenarios than carbon-based life. In support of this idea is also that whereas hundreds of organic molecules based on carbon chemistry have been detected in outer space[11], it has not been so with similar molecules based on silicon, perhaps for the reasons explained above, especially the need for oxygen being absent to allow the formation of silicon chains. Moreover, we are not talking here only about simple life, but about its evolution towards complex enough organisms that could reach the point of forming a civilization. Given the above difficulties, this is even more unlikely with a life based on silicon chemistry, or developing in a non-aqueous media, even if it were possible. Life, and especially civilizations elsewhere in the universe, must also be based on organisms formed by carbon-chain molecules. For this reason, the planets on which life and civilizations may develop must possess the pressure and temperature conditions allowing the presence of liquid water. These requirements drastically reduce the number of planets on which life could emerge and evolve to civilization because, among other conditions, the planet shall be located at appropriate distances from it star so that these requirements are met. We will discuss this issue again later, but let's talk now about the evolution of intelligence on our planet, Earth.

Notes to chapter 2

1 http://www.ncbi.nlm.nih.gov/pmc/articles/PMC516796/?tool=pubmed. -
http://www.una.edu/faculty/pgdavison/BI 101/Overview Fall 2004.htm -
http://www.ncbi.nlm.nih.gov/pubmed/11312589

2 http://www.galleries.com/minerals/SILICATE/class.htm

3 http://www.worldcat.org/title/supernovae-and-nucleosynthesis-an-investigation-
of-the-history-of-matter-from-the-big-bang-to-the-present/oclc/33162440 -
http://en.wikipedia.org/wiki/Abundance_of_the_chemical_elements -
http://www.chemeurope.com/lexikon/e/Abundance_of_the_chemical_elements/

4 http://www.space.com/supernovas/

5 http://en.wikipedia.org/wiki/Abundances_of_the_elements_(data_page)

6 http://www.entropylaw.com/

7 http://www.statemaster.com/encyclopedia/Spontaneous-process -
http://en.wikipedia.org/wiki/Spontaneous_process

8 http://www.absoluteastronomy.com/topics/Vitalism -
http://en.wikipedia.org/wiki/Vitalism

9 http://www1.lsbu.ac.uk/water/molecule.html -
http://www.chem1.com/acad/sci/aboutwater.html -
http://en.wikipedia.org/wiki/Water_molecule#Hydrogen_bonding

10 http://www.daviddarling.info/encyclopedia/A/ammonialife.html

11 http://hypography.com/news/astronomy/35483.html

Chapter 3: Intelligences

The development of life on any planet would be useless if it did not evolve long enough to originate intelligence, which I consider absolutely necessary for a technologically advanced civilization, capable of communicating with others, to arise in due time. The evolution of life towards intelligence does not seem an easy process, considering that on our planet life took at least three billion years since its origin until the appearance of the first organisms to which some kind of primitive intelligence may be attributed[1].

As we did with life, we should define what intelligence is before we talk about it. However, in my opinion, we also lack an intelligent definition of intelligence. Perhaps this deficit is due to an excessive emphasis on defining human intelligence, not intelligence in general. In our case, it is important to define intelligence broadly, i.e. including the kind the simplest worm may possess.

In any case, we can find many definitions of intelligence in the literature and the Internet. One of the simpler ones, found in Wikipedia, is: the ability to apply knowledge to perform better in a given environment[2]. This definition implies, first, that intelligent organisms must possess the ability to acquire knowledge from the environment in which they live. In turn, this raises the difficulty of defining what knowledge is, a subject on which many works related to philosophy, psychology or neuroscience have been written.

As far as I'm concerned, I prefer to define knowledge as the ability to internally simulate the external environment.

That is, one knows when is able to conceptualize objects and simulate processes that exist or happen in the external reality of the organism "imagining" them. Of course, the internal simulation process is impossible if the intelligent organism is not capable of drawing generalizations about the environment in which it lives. The imagination and simulation on such generalizations can anticipate the contingencies that may occur, and allow the organisms to adapt and perform better in the environment.

From the strict point of view of survival in a given environment, there are two main kinds of intelligence, in my opinion: 1. Intelligence allowing adapting to changes in the environment. 2. Intelligence allowing using the environment and adapting it to the advantage of oneself. This second type of intelligence not only allows better adaptation to the environment, but also to make environments adapted to the living organism, and even allows creating new environments that best meet the needs of the organisms creating them. Clearly, though many animals are capable of performing minor changes in their environment (building nests, tunnels, or a variety of shelters, for example), only one species is intelligent enough to possess the second kind of fully developed intelligence (or almost), as to govern Nature by obeying, and above all, understanding its laws. This species is we, the humans.

However, to make possible any modification of the environment it is not only necessary to possess a lot of intelligence; other skills are also required, such as the ability to act on the external environment. This implies that the ability to change the environment also depends on the anatomic features of the living organisms. A technological civilization emerging from a society of intelligent snakes, which only could crawl on the floor, occasionally pulling their forked tongue out with the intention of manipulating something would be hard to imagine. Fortunately, throughout its evolution, our species has developed hands. The hands were developed by our ancestors because they were necessary, or beneficial, to adapt to an arboreal environment, so that they could hold on to the tree branches, and thus perhaps better escape predators. The hands allow us now, along with intelligence, to change our environment. Although our species is not alone in possessing hands, and many arboreal animals also possess them, it is unique in having the right combination of intelligence and manipulative ability to substantially alter the environment, a necessary condition to develop a technological civilization.

The appearance of hands during the evolution of land animals does not appear to be particularly difficult, though, in my view, the evolution of sophisticated hands, like ours, depends first on the evolution of land plants. Only when these have evolved in the history of life to the point in which their constitution is strong enough to withstand animals of a certain weight, an ecological niche could be created allowing the development of hands with the purpose of holding and moving in the arboreal environment. However, there seems to be no particular reason that would prevent the evolution of land plants, once they appeared in evolutionary history, toward bigger and stronger structures, such as those typical of the trees. The ecological niche for the evolution of arboreal animals will be there once trees have appeared and, therefore, it will likely lead to the evolution of hands.

However, what does not seem so easy is the emergence of land plants, necessary as I said to first enable the emergence of terrestrial animals that could develop hands. Moreover, the difficulty for the appearance in the history of evolution of the second type of intelligence, that which allows modifying the environment, is even greater. We will go inside some of the reasons why it is not easy that neither land plants nor this kind of intelligence appear in the history of life.

First, we must remember that once the first living organisms appeared, the first bacteria, they reigned over the earth in solitude for about two billion years[3]. Let us pause here for a moment, two billion years deserve it. It is estimated that the split from the common ancestor of chimpanzee and human species occurred only five to seven million years ago[4]. That is, our own evolution as a species has occurred in only around a 0.0025% of the time that bacteria reigned alone over the Earth.

Bacteria and other prokaryotic organisms have not disappeared from our planet, but today they are not alone. They are accompanied by millions of others organisms, more complex than them. These organisms are the result of evolutionary events occurring to bacteria, from which they, and we, derive.

Why bacteria took two billion years to evolve towards other organisms? Nobody knows for sure, but what seems certain is that if it took so long, the event which gave rise to more complex organisms, which led to the first eukaryotic cells with a cell nucleus, then allowing the development of multicellular animals and plants, was not very likely. To understand why, consider the time we should be tossing a coin to get one hundred heads in a row. Since this event is extremely unlikely, although not impossible, we should be throwing again and again the coin for a long time, without ever having the security of getting the result we seek. In any case, the event that allowed the evolution of bacteria to more complex organisms finally happened. This event seems to be, according to the latest research, the fusion of two organisms of different and complementary types, which entered into a symbiosis so close that the two organisms eventually merged into one[5]. This fusion gave rise to eukaryotic cells, from which all multicellular organisms evolved. The concept of symbiosis appears here again. Symbiosis is, in my opinion, the core of life on several planes, from the molecular level to the organisms themselves.

This is important for the topic at hand, because it does not appear possible that a civilization might arise on any planet without the prior emergence of complex multicellular organisms, the only ones with the capability to manipulate the external environment. Undoubtedly, a single cell can do little to manipulate anything. The origin of eukaryotic cells is also important because they acquired a much higher capacity than bacteria to obtain energy from the environment, thanks to mitochondria and chloroplasts. This higher capacity to obtain energy is essential to obtain and maintain new genes. The reason is that genes are DNA molecules that replicate with a high cost in energy. Prokaryotic organisms cannot get much energy from the environment, so they tend to lose all those genes that are not strictly necessary for their life and reproduction. Therefore, prokaryotes have an energy limit for increasing the complexity of their genomes. However, eukaryotic cells can afford to maintain and replicate new genes acquired through various means, including gene duplication or transposition of genes from other organisms[6]. This capability is very important for evolution, because the more the genes that can mutate, the better for the generation of new organisms that can be selected by natural selection.

That may be the reason why multicellular organisms are composed exclusively of eukaryotic cells. Only this more complex type of cell was able to evolve and acquire the genes necessary to enable it to collaborate with similar cells, and above all, to gain the ability called cell differentiation. Cell differentiation is the process by which a stem cell can give rise to different specialized cells, each of them exerting a different function in the body. Thus, from a single initial cell, different types of cells originate and get organized in various organs and tissues with special functions, such as the liver, the intestine, the kidney and the brain. The ability of cell differentiation is specific of eukaryotic cells, and it would be impossible if they did not possess an extensive collection of genes. The reason is that the differences between the several types of differentiated cells arise because different set of genes is operating in them at a given time. That is, the genes that are active in a nerve cell are not the same that those operating in, for instance, an epithelial cell, although both cells have identical sets of genes in their chromosomes.

However, it took a long time, again, for the emergence of the first multicellular organisms from unicellular eukaryotic cells. As with the first appearance of eukaryotes from prokaryotes, the emergence of multicellular organisms from unicellular eukaryotes was an unlikely process. In fact, it took between one and one a half billion years since the appearance of the first eukaryotic cell. That is, this event happened about 630 to 580 million years ago, when the so-called Ediacara biota[7] emerged. This occurred in an era before the Cambrian geological period, which began 542 million years ago, a time where a great deal of biodiversity emerged, called the Cambrian explosion, which leaded to the appearance of the first cephalopods[8].

After about 3.5 billion years since the appearance of life, we finally encounter multicellular organisms, the only ones capable to continue evolving towards intelligent beings. But, so far, the entire evolution of life has taken place in the ocean. None of these organisms has colonized dry land yet. And the colonization of land by living organisms seems vital for the development of intelligence, enabling, in turn, the development of a technological civilization. This is an important point that can help us resolve the Fermi paradox. Here is why.

Land and intelligence

First, it is important to consider that the most intelligent marine animal is a cephalopod, probably the octopus[9]. However, you may think that's not true. No doubt dolphins and porpoises, killer whales and even the clumsiest and biggest whales are more intelligent than the octopus. You're right. The problem is that these animals are not strictly marine ones. Marine mammals are animals that, after passing a time of their evolution on mainland, got adapted again to living in an aqueous environment[10]. They were originally land mammals for the simple reason that mammals appeared on the mainland[11]. So, we should not view the dolphins and their brothers and cousins as pure marine animals. This leaves us again with the octopus as the smarter sea animal. What does this mean?

Well, in my opinion, this means that the kind of intelligence capable of developing a technological civilization cannot appear in the ocean. If water is absolutely necessary for life to emerge, and as we said, necessary to maintain the order and complexity proper of life, mainland is necessary for the development of truly intelligent organisms. This dichotomy between water and non-water, but at a different level, reappears again as a necessary condition, if not for life this time, for the emergence of intelligence beings capable of developing a civilization

What evidence do we have to argue in favor of this idea?

The first plants on the mainland date back 475 million years. Only after that time the animals were able to conquer the mainland independently of the aqueous medium, i.e. they were able to get food from a different source than the sea. This happened during the Devonian period, 410 to 354 million years ago[12]. As we said, complex organisms appeared in the ocean 630 to 580 million years ago. That is, these organisms were about 200 million years ahead of the most primitive terrestrial organisms to evolve higher intelligence; however, this did not happen. Nevertheless, probably the first smart animals, the reptiles, some of them possessing at that time perhaps a level of intelligence comparable to that of our dogs, appeared on the mainland around 320 million years ago, i.e. only 170 million years after the colonization of mainland by plants, which was followed by that of the animals[13]. Thus, a high level of intelligence appeared on land in less time than that marine animals had to evolve intelligence at sea before conquering the land, despite which they did not evolve intelligence at a higher level than that shown by the octopus, not even until today. This suggests that well-developed intelligence does not easily appear and evolve in the ocean.

There are some important reasons to explain why this is so. The most important, in my view, is that the marine environment offers much smoother and predictable conditions for life that the terrestrial environment. For example, sudden temperature changes do not happen in the ocean. There are no storms; the seasons do not exert their periodic effect on climate so dramatically as on land, because, in fact, there is no climate. It does not rain, nor are there any droughts. It never freezes, nor the sun is ever burning too much. In short, the marine environment is much more constant and predictable than the terrestrial environment. To survive, the need to develop complex simulation mechanisms and generalizations (a complex nervous system) to adapt to this environment is much less than the need to do so in the terrestrial environment. This is probably the main reason why most intelligent marine animals are actually terrestrial animals reconverted to sea animals, and also the reason explaining why a large number of the most intelligent species are terrestrial, like us and our cousins, the primates, to name a few.

Besides, the aqueous medium imposes certain restrictions on the animals that live in it. With the exception, precisely, of the octopus and other cephalopods, limbs are conspicuously absent in marine animals. They must acquire a hydrodynamic shape allowing them to move economically through the water. The absence of limbs in marine animals is so notorious that even the animals that once had acquired them during their mainland evolution time, transformed them into flippers to adapt to sea life again, as it happened with cetaceans.

Without limbs it is practically impossible to manipulate the environment in which we live. Consider what we would become if we had not our dear hands. Losing both hands is possibly one of the greatest misfortunes that, along with blindness, a human being can suffer. Yet, some animals have no hands all their lives, especially the marine ones.

It's hard to imagine that any kind of crippled animals, like the snakes I mentioned before, whatever smart they were, could develop a technologically advanced civilization, able to adapt their environment to their needs. However, even when intelligent animals, able to develop manipulative organs, may exist on land, their existence do not necessarily implies civilization will develop. For example, the dinosaurs dominated the Earth for 160 million years or more[14], but along that time, although some were bipedal and had rudimentary hands, no species of dinosaurs, that we know it, raised to develop a technological civilization. However, only five to seven million years after the appearance of the common ancestor of chimpanzees and humans, civilization arose. Therefore, we must conclude that not on all planets with evolutionary circumstances similar to ours, technologically advanced civilizations will arise whenever possible to do so. Perhaps the dinosaurs did not have the selection pressure from other more evolved species competing with them, as was the case with our ancestors, favoring the development of superior intelligence as a tool for survival and the transmission of genes. That is, perhaps life must evolve beyond a certain point on the mainland to favor the development of intelligence. However, this development might never occur.

In addition to the anatomical constraints faced by marine animals, there are other equally important constraints that hinder the development of a technologically advanced civilization in the marine environment. Perhaps the most important of them is that we cannot light a fire into water. The control of fire was undoubtedly one of the milestones of human development[15]. Without fire, and generally without the ability to generate energy in a relatively simple way, technological civilizations would be impossible. This ability to generate power in a simple manner occurs only on the mainland, but it is much harder under water. This, together with the fact that purely marine animals are less intelligent than terrestrial ones, greatly hinder the development of underwater energy technologies on other planets, if they were possible. If humans are currently able to use devices that produce energy or release it under water, is thanks to that the technological development of these devices has been carried out on land. In other words, a submarine, or a bathyscaphe, or any ship or boat, are only possible, paradoxically, because the technology that makes them possible has been developed on the mainland. It is a situation similar to that found with the intelligence of dolphins, porpoises, seals and whales, of which we spoke before.

Before we finish discussing this issue, it should be mentioned that the ability to light a fire on the mainland depends on the presence of an oxygen-rich atmosphere. This gas can only accumulate in the atmosphere, as far as we know, as a result of a process proper of life, which, as we clarified before, can only develop in an aqueous media. This process is the photosynthesis[16]. This means that for the appearance of a species able to use a simple energy source, such as combustion, life needs to have "discovered" photosynthesis, and carried it out for hundreds of millions of years to allow sufficient accumulation of oxygen in the atmosphere. In addition, burning materials are either alive now (plants and trees, for instance) or were alive before (wood or fossil fuels). Inorganic material does not burn under normal conditions. That is, the capacity to generate power through fire depends absolutely on the continual activity of living beings on our planet. However, the presence of significant amounts of oxygen in the Earth's atmosphere did not guarantee the emergence of sufficiently intelligent organisms as to develop technology.

To summarize the two previous chapters and this one, we can say that life unfolding in a distant planet orbiting a star far away will be based on carbon. Moreover, for life to develop, the planet must hold abundant liquid water, which limits the range of distances at which the planet can be located from its star. But the planet should not have so much water as to completely cover its surface, because then the development of life on the mainland will be impossible. This will significantly impair the development of organisms smart enough to generate and use energy, and possessing limbs with which to manipulate the world and eventually develop a civilization. Let us now finally analyze how the Moon, our dear satellite, which in Spain accompanies the bulls in love (according to a well-known popular song), was able to exert a fundamental role facilitating the conquest of land by living organisms.

Notes to chapter 3

1 http://www.sciencemag.org/cgi/content/summary/300/5626/1691 -
http://exploringorigins.org/timeline.html -
http://en.wikipedia.org/wiki/Evolutionary_history_of_life

2 http://en.wikipedia.org/wiki/Intelligence

3 http://exploringorigins.org/timeline.html

4 http://www.newscientist.com/movie/becoming-human -
http://en.wikipedia.org/wiki/Human_evolution

5 http://www.tolweb.org/Eukaryotes/3

6 Nick Lane. Power, Sex, Suicide: Mitochondria and the Meaning of Life. Oxford
University Press, USA (2006).

7 http://cambrian.tripod.com/IntrotoEdiacaran -
http://www.britannica.com/EBchecked/topic/179126/Ediacara-fauna -
http://en.wikipedia.org/wiki/Eukaryotic#Origin_and_evolution

8 http://www.paleo.pan.pl/people/Dzik/Publications/Cephalopoda.pdf -
http://www3.interscience.wiley.com/cgi-bin/bookhome/117345722/ -
http://www.plosone.org/article/info%3Adoi%2F10.1371%2Fjournal.pone.0007262

9 http://discovermagazine.com/2003/oct/featmeye -http://www.slate.com/id/2192211/

10 http://www.britannica.com/EBchecked/topic/103892/cetacean -
http://en.wikipedia.org/wiki/Sea_mammals

11 T.S. Kemp. The Origin and Evolution of Mammals. Oxford University Press, USA
(2005). ISBN-10: 0198507615. ISBN-13: 978-0198507611 -
http://www.bobpickett.org/evolution_of_mammals.htm -
http://en.wikipedia.org/wiki/Mammals#Evolutionary_history

12 http://www.ucmp.berkeley.edu/devonian/devonian.html

13 http://en.wikipedia.org/wiki/Evolutionary_history_of_life

14 http://www.nhm.ac.uk/jdsml/nature-online/dino-
directory/timeline.dsml?disp=gall&per_id=&sort=Genus

15 http://en.wikipedia.org/wiki/Fire#Human_control

16 http://www.life.illinois.edu/govindjee/paper/gov.html
http://www.estrellamountain.edu/faculty/farabee/BIOBK/BioBookPS.html -
http://biology.clc.uc.edu/Courses/Bio104/photosyn.htm

Chapter 4: Moons

As mentioned briefly before, one of the arguments put forward in favor of the importance of our natural satellite, the Moon, on the evolution of life on Earth, is its gravitational effect, causing the tides. The writer and scientist Isaac Asimov, in his book entitled "The Tragedy of the Moon"[1], published in 1971, pointed out that life, emerging in the ocean, and therefore evolving and adapting to the marine environment, had no reason to colonize the mainland, had not been helped, or forced to do so, by receiving a "little push" out of the water. That "little push" could well have been given, at a distance, by the tides caused by the Moon.

As we know, the tides are periodic elevations of the sea surface caused by the combined gravitational pull of the Moon and the Sun. The explanation of the phenomenon in gravitational terms is relatively simple. Earth rotates on its axis once every 24 hours. This causes, at any given time, a portion of its surface to be located closer to the Moon than the diametrically opposite side of the planet. Since the gravitational attraction depends on the distance, the side of Earth closest to the Moon receives a greater gravitational pull than the far side. This pull slightly deforms the Earth, which acquires a more "egg-shaped" form towards the Moon. The deformation of the Earth's surface happens everywhere, but is more intense on the ocean, as it is fluid and can be deformed more easily than the solid Earth's crust. Since the Earth rotates around its axis much faster than the Moon revolves around the Earth, this distortion is continuously moving "backwards" on the Earth's surface as the Earth rotates and presents a different part of its surface to the Moon. A similar effect is exerted by the gravitational pull of the Sun, which can be added or subtracted to the Moon's, according to their positions relative to Earth. Thus, if the Moon, Earth and Sun are placed in a straight line, which happens about twice a month, the gravitational pulls of the Moon and Sun are added, and tides are higher. If, however, Moon, Earth and Sun are at right angles, the gravitational pulls of the Moon and Sun are partly offset, and tides are lower.

Tides cause, therefore, a rise and fall of the ocean's and sea's surface, with peaks of ups and downs about once every six hours. In addition to the relative positions of Earth, Moon and Sun, the intensity of tides depends on other factors, such as the latitude or the particular coastal features. In coastal areas with a gentle slope, water can invade large areas of land. In areas where the slope is steeper, the invasion of land by the sea is smaller[2].

Tides originate, then, an ecological niche linking the land and the sea. In this area modern organisms, both animals and plants, live today adapted to the sudden changes in the environment produced by this periodic rise and fall of the water levels. In his book, Isaac Asimov argued that, without the tides, a stimulating environment for the conquest of land by marine organisms would have not existed. Lacking this intermediate environment, living organisms could not have developed a skin able to protect them from drying out, limbs to move, and in short, physiological mechanisms and processes to survive outside the aqueous environment. Without tides, Asimov argued quite convincingly, multicellular organisms of a certain complexity would had never left the ocean, at least they would have not done so quickly, and, I add, the evolution of intelligent beings would had not happened, since, as we have already explained, well-developed intelligence probably only arises during the evolution of terrestrial beings.

However, maybe this is claiming too much. It is possible that, for other reasons, life could have colonized land anyway. To start with, tides are not just caused by the Moon, but also, as we said, by the Sun. In addition, it is even conceivable that, even without any tides, certain coastal organisms could have been driven to colonize land by other reasons, such as escaping from a marine predator, for example, or simply to benefit from a greater flow of sunlight to carry out photosynthesis, because water shields solar radiation. Maybe those animals or plants capable of living on land for a short while could survive better as a result. In sum, the Moon is not strictly necessary for life to colonize land. However, it is clear that its gravitational effect, causing higher tides, could have significantly accelerated this process. That is, a planet with a moon, also meeting the conditions necessary for life to emerge and evolve, may possess a greater likelihood for the development on its continents of intelligent beings, capable of creating civilizations long before other planets with similar conditions, but lacking a moon.

To understand this better we should halt here for a while and explain what I define as an "evolutionary space-time". This concept should be differentiated from that used in evolutionary theory, called "fitness landscape"[3]. In my view, a given evolutionary space-time should be defined by a set of environmental conditions, compatible with life but changing over time so that the space would *favor the evolution of organisms in a particular direction.* In other words, an evolutionary space-time is not constituted by any static set of environmental conditions, but by a set of conditions that gradually or periodically evolve so as to permit, in turn, the evolution of organisms in a particular direction.

Let's try to illustrate this with an example. If a group of researchers intended to generate a new breed of laboratory flies more resistant to drought, or adapted to the desert, what they would not do would be to place a population of flies in a completely dry environment, with no water at all, let the flies to evolve and adapt, and select the survivors: there would be none. Without any water, all the flies from the initial population would die. Therefore, what it should be done would be placing the initial population of flies in an environment with slightly reduced water availability. This reduced water availability, which must anyhow allow the flies to continue being alive, would create a "selection pressure", i.e. it would allow those flies that by any chance were more resistant to water scarcity to reproduce better than the less resistant flies and, therefore, transmit their genes to the next generation with a greater probability. After this initial selection, the researchers would decrease the availability of water a little further, to select in this way flies still more resistant to water scarcity. Thus, generation after generation of flies, by gradually decreasing water availability, the researchers would select those individuals whose genes allow them to survive better in the desert. Obviously, using this evolutionary strategy we can only modify the artificial conditions of the environment to a certain a limit, which in this case would be the least amount of water compatible with keeping the flies alive. Below that amount no further evolution would be possible, because there would be no life. We would have reached what is called an ecological crisis, leading to extinction. That is, life can only evolve within a set of conditions of temperature, humidity, etc., allowing its existence. But to favor life evolving in a particular direction, these conditions must change constantly and progressively so that their variation continues to be compatible with the life of the initial organisms. Since it is inevitable that mutations gradually occur, their genes will vary, thus allowing natural selection of mutants capable of reproducing more successfully in the evolving environment.

The idea of constant change and in a given trend is important. Obviously, the scientists mentioned above will not be able to select a race of flies resistant to drier conditions if at each generation they randomly vary the amount of water available to the flies, so that sometimes the flies have less water, but other times they have more. This random variation of the conditions happens every year in Nature, with drier or wetter, colder or warmer seasons. The natural evolutionary space-times tend to vary in a non-directed way, i.e. with no specific trend. However, if conditions change gradually, organisms will suffer progressive adaptations that may enable them to live in a niche in which they could not live before. For example, glaciations might help the emergence of species adapted to live in polar climates, like the arctic fox and the polar bear. Perhaps, in the absence of these glaciations, the ancestors of these animals would not have been selected to survive better in cold climates and polar bears and arctic foxes would not exist today. In fact, polar bears appear to have derived from Grizzly bears less than 200,000 years ago, from a population of these that became isolated due to the Pleistocene glacial period[4].

Well, the Moon, with its tides, put in place an evolutionary space-time that has accelerated the evolution of animals and plants towards varieties capable of surviving outside the ocean. When multicellular marine plants had already colonized the coasts of seas and oceans, at least 490 million years ago, the tides at that epoch created an environment in which sea water periodically flooded and then leaved some coastal areas for a short period of time, maybe insufficient in many cases, to kill most marine plants that may have begun to grow in those tidal areas. However, the period of absence of sea water created a "selection pressure" favoring the survival of living beings more resistant to water shortages. In addition, and very importantly, not all parts of the tidal zones, i.e. not all the areas periodically flooded by seawater, remain non submerged the same length of time. The areas closest to the sea are initially flooded early by the tide and continue flooded almost the entire duration of the tidal period. But tidal areas farther from the coast are inundated only for a short time, just at the time of the high tide. From this moment, the tide gets lower and seawater abandon these points. That is, from the sea to the high tide point an area gradient is created where seawater is present or absent for increasing amounts of time. The areas near the sea are submerged, or wet, for a long time, but the most far away areas are submerged and wet very little time. Between these two points, depending on their distance from the sea, the tidal zone is submerged or dry for periods of time continuously increasing or decreasing, depending of the point of view.

It is easy to see that thus tides create a continuous and homogeneous evolutionary space-time, which might have directed the evolution of multicellular organisms towards adapting to life on land. We can imagine, and this is just my presumption at this point, that initially some algae were deposited, carried by storm waves, in an area outside the sea but close to it; wetted only by large waves and by the tides. Being close to the sea, this area was submerged for long periods of time although for short periods it was not under water. Under these conditions, algae capable of withstanding this short period out of water, survived.

The algae that survived became perhaps the ancestors of plants able to settle a little further from the coast, at points where the tide could not keep them submerged for so long. That is, in a short space, only the tidal zones, thanks mainly to the Moon (and also, although in much a lesser extent, to the Sun) a continuum evolutionary space-time was generated, in which distance from the sea determined the conditions of humidity and dryness. This evolutionary space-time could thus be colonized by variants of algae or other marine plants adapted to grow at increasing distances from the sea, according to their ability to withstand the absence of water. It's easy to think that when the algae got adapted to live in areas far away from the sea, but still wetted by the tides, they could withstand the lack of seawater for a long period of time. These plants could take the final evolutionary step and start living only on land areas flooded by rain. Of course, this process took millions of years. Obviously, a similar process could enable the animals to venture out of the marine environment and begin to colonize the continents of the world.

Thus, the Moon, causing heavy tides, which were heavier, as we shall see, a few hundred million years ago, when the entire drama of terrestrial life evolution began to unfold, has played a major role in generation of an evolutionary space-time leading to the conquest of land by marine life. We have already explained that, had this conquest not happened, organisms smart enough to develop a technological civilization probably would have not developed. So, elsewhere in the universe, even on other planets just like Earth, but with no Moon, this conquest of the land either would have not occurred, or would have happened more slowly than it did on our planet.

Planet frequency

We now may ask: what is the frequency of planets like ours having a moon like ours? The number of planets on which life might have originated and perhaps evolved towards a technological civilization may depend on the answer to this question.

To answer this question, we must first answer the question whether or not planetary systems like ours are common around stars. At present, it is premature to answer this with certainty. However, as I write (February, 2011), 526 planets orbiting other stars have been cataloged, according to information documented on the extrasolar planet encyclopedia[5]. This number is not bad at all, is capable of allowing some statistical analysis of the characteristics of the new planets and solar systems discovered, and start estimating the likelihood of existence of other terrestrial planets. It is also expected that the Kepler telescope[6], launched in March 2009, will continue to provide valuable information on the percentage of stars with a cohort of planets, and about the proportion of those planets that may possess the necessary conditions for the development of life. According to the latest data from the Kepler mission, made public in February 2011, Earth-like planets are not very abundant. In the coming years, the number of Earth-like planets that could exist in the galaxy will be estimated with some degree of accuracy, although it will be much more difficult to estimate whether or not they have moons similar to ours.

However, according to the data already available, it is estimated that there are more planets than stars in the universe[7]. This is good news for life. However, most planets discovered so far are gas giants, even more massive than our planet Jupiter. Moreover, unlike what happens with our Solar System, these massive planets orbit very close to its star, occupying the space in which Earth-like planets might be orbiting[8]. The planetary distribution of our Solar System, with rocky planets orbiting near the star and gas planet doing it far away seems to be the exception, rather than the rule. Formerly it was reasoned that it was logical to expect a planetary distribution similar to ours in other planetary systems. The reasoning was based on the assumption that near the central star a massive core could not form and compete to drawing and collecting matter around the area, so that giant planets might have formed only away from the star.

This assumption has proved completely false, since astronomers have discovered planets more massive than Jupiter orbiting stars at shorter distances than Mercury is from the Sun. However, it is true that these planets initially form far away from the star to migrate and stabilize their orbits later very near the star[9].

This feature may be important for the development of life in other solar systems. For life to develop on a planet, this must be located into what is called the habitable zone[10]. The presence of giant planets orbiting near the star could, in some cases, make this impossible for small rocky planets like Earth, likely the only kind of planets capable of holding liquid water at their surface. The habitable zone around a star is defined, by the way, as the one in which the temperature is suitable for liquid water to exist. We have already explained in Chapter 2 that it is unlikely that life could exist in the absence of liquid water, and without the molecular complexity that carbon chemistry makes possible. In the case of the Sun, its habitable zone is estimated to extend from 0.95 to 1.37 times its distance from Earth, according to the latest models, although others may shorten or lengthen this distance[11]. Obviously, Earth is placed on the distance value 1.0, which is called, in technical language, an astronomical unit: simply, the distance from Earth to the Sun.

Some scientists have speculated with the idea that the formation of giant planets and their migration to a close orbit around their stars could facilitate the formation of Earth-like rocky planets in the habitable zone. In addition, these giant planets may possess rocky satellites, as it happens with Jupiter or Saturn, in which life could develop. It could even happen that a giant planet and its many satellites orbiting inside the habitable zone of its star had several of them where life could flourish. However, they should be of considerable size and mass, at least above the size of Mars, otherwise they would not possess enough gravity to keep liquid water on its surface by gravitational attraction. This is what happens with the Moon, which in spite of orbiting with us in the habitable zone, contains practically no water and, therefore, no life. It also happens to planet Mars, located at the outer edge of the habitable zone. Although Mars held once an ocean of water on its surface, now it holds almost none because its surface gravity (only 0.38 times that of Earth) is not sufficiently great to retain it.

We must clarify that the fact that most of the planets discovered so far are very large, even exceeding the size of Jupiter, is not due to these planets being more numerous, but because it is easier to detect large planets and more difficult to detect small ones, similar to Earth. In fact, many planetary scientists believe that the formation of Jupiter-size or greater planets is more difficult than that of smaller rocky planets, like Earth[12]. This means that for every star with a planet like Jupiter, it is reasonable to expect one or more planets like Earth orbiting around the star as well. In fact, in our Solar System for each gas planet (Jupiter, Saturn, Uranus and Neptune), we have, indeed, one rocky planet (Mercury, Venus, Earth and Mars, apart from the asteroid belt, where another rocky planet should be located, had Jupiter's gravity not prevented it[13]).

Pending what the Kepler telescope and other observational instruments, such as COROT, which discovered the first rocky planet around a star[14], will finally reveal to us, we may therefore presume that each star in the galaxy similar to our Sun possesses at least one Earth-like planet. There are an estimated one hundred billion stars like the Sun in the galaxy, so if this assumption is correct, there would be one hundred billion Earth-like planets revolving around them. Given this huge number of planets similar to Earth one can ask again, along with Fermi: why haven't we been visited by alien civilizations yet?

The reason may well be that, for the development of civilizations, not only Earth-like planets are required, but Earth-like planets with satellites like our Moon. What is the expected abundance of moons similar to ours around planets like Earth?

Moon abundance

To answer this question we need to know how our Moon was formed, and what is the probability of moon-like satellite formation around planets similar to ours. Let's visit the different scenarios proposed on the origin of the Moon and analyze, in light of what we now know about our natural satellite, which theory, if any, fits reality.

Obviously, scientific hypotheses on the origin of the Moon had to await the birth of astronomy, and only after Galileo turned his telescope at Jupiter and discovered that this planet also has satellites, astronomers began thinking about the origin of the Moon. However, the first person who proposed a hypothesis on the origin of the Moon is not exactly considered an astronomer, but a philosopher and mathematician, and one of the most influential too: René Descartes. Descartes, while thinking that he was thinking, therefore he was, also thought the idea that the Moon, following a close approach, was captured by the Earth's gravity, and was placed in orbit around it. This idea has been called the capture hypothesis. In view of the problems Galileo had with the Church, Descartes, a fine thinker about existence, but also about survival, did not publish his hypothesis while alive, which if did not guarantee him a longer existence, it sure did a more peaceful one. Descartes's hypothesis was published in 1664, fourteen years after his death[15], a time long enough to make sure he was really dead. A modern version of this hypothesis had to wait till nothing less than 1909, when the American astronomer Thomas Jefferson Jackson See[16] suggested that the Moon was originally a planet orbiting the Sun, whose orbit led to its capture by the Earth.

The second hypothesis on the origin of the Moon was proposed, in 1878, by none other than George Howard Darwin, son of the famous Charles Darwin, one of the fathers of the theory of Evolution. This son of Darwin proposed that long ago the Earth revolved so fast and was so egg-shaped by that reason that, aided by the gravitational pull from the Sun, it lost a large chunk of its surface, which became the Moon. Four years later, this hypothesis was reinforced by another one, proposed by the geologist Osmond Fisher[17]: the Pacific Ocean was the scar left by the loss of the Moon, clear evidence, he claimed, that Darwin's hypothesis was true[18]. This idea was called the fission hypothesis.

The third hypothesis on the origin of the Moon claims that it was formed at the same time as the Earth by accretion of the dust particles that initially revolved around the central mass of the accretion disk giving rise to the Sun and planets.

This hypothesis was put forward by the astronomer Edouard Roche[19], among others, and was called the hypothesis of co-accretion, or co-formation.

The fourth and final hypothesis on the origin of the Moon is the collision hypothesis. This hypothesis, proposed by Reginald Aldworth Daly, of Harvard University, in the decade of the 40's of the last century[20], claims that in the early Solar System, a high mass planetoid collided with Earth only about 40 million years after its initial formation[21]. As a result of this massive collision, part of the mass of Earth and the planetoid was ejected to outer space, where it was placed in orbit around the Earth. In just 100 years, this matter coalesced by gravity and formed the Moon[22].

Which of these hypotheses is correct, if ever any of them is? To find out, it has been necessary to conduct a long scientific process, probably impossible to carry out without the Apollo and Russian missions to the Moon, which altogether brought to Earth about 400 kilograms (900 pounds) of lunar rocks for analysis. We will go reeling off briefly this process, because in addition to learning more about the Moon, it will allow us to see how science works and how it makes possible the progress of knowledge for mankind.

How is our Moon?

To begin with, let's make a tour along the features of the Moon that astronomical and geochemical studies, among others, have revealed. Any hypothesis aiming to explain its origin must necessarily be compatible with these features. For example, if we find that the Moon and Earth have exactly the same chemical composition, the capture hypothesis would be in doubt, since it is unlikely that two bodies formed in different parts of the Solar System should have an identical chemical composition. This is just one example of how to proceed in reinforcing or disproving a given hypothesis. Let's see the peculiarities of the Moon and which ones reinforce or refute the previous hypotheses.

It turns out that the Moon is a satellite rather strange, as compared to other satellites in the Solar System. The first oddity of the Moon is that its iron core is very small. This is extremely rare, since iron is an element formed abundantly in supernovae stars, one of which exploded and allowed the formation of the Solar System, as explained. In fact, iron is the sixth most abundant chemical element in the universe, and also the sixth in abundance in the Solar System[23]; that is, even more abundant than silicon, which, let's recall it, is the eighth[24]. To this we must add that the chemical composition of lunar rocks is very similar to those of Earth, in particular, the analysis of the oxygen isotope ratio indicates that this ratio is identical in Moon and Earth, which implies that the matter that formed the Moon had to come from the same source as that which formed the Earth; that is, the same supernova that gave birth to the Solar System. But not only that: the isotopic composition (particularly of oxygen and xenon) indicates that the Moon and the Earth must have been formed at the same region of the Solar System, since different regions of it do not possess the same chemical and isotopic composition than that found on our planet, as demonstrated by the analysis of meteorites and asteroids fallen on Earth[25]. The above means that the Moon should possess an iron core proportionally similar in size to that of Earth, and for this reason the deficiency in iron content of the Moon is even more mysterious. Iron, being very heavy and dense, ends up in the core of rocky planets and satellites during their formation by gravitational effects. However, whereas the Earth has an iron core accounting for 30% of its mass, that of the Moon only accounts for 3% of it[26]. In fact, the estimate of the iron and silicon ratio on the Moon reveals that it is the smallest of the entire Solar System. That is, the Moon is the body of the Solar System possessing less iron in proportion to silicon. In addition, the Moon also possesses the lowest proportion of heavy elements less common than iron. Why is this so, if other evidence suggests that Earth and Moon formed at the same region of the Solar System? However, an interesting difference between Earth and Moon volcanic rocks is the lack of water in the latter's. In general, the Moon possesses in the upper mantle a smaller proportion of light elements than the Earth [27]. It should not be like this for two bodies forming at the same region of the Solar System, and at the same time. How can we resolve this dilemma? Certainly, it cannot be explained by differences in the gravitational force exerted by Moon and Earth, as Io, a satellite of Jupiter similar in size to the Moon, does possess light elements in greater proportion. Clearly, any hypothesis about the origin of the Moon should be able to explain this discrepancy.

To increase the mystery even further, the study of Moon's rock samples shows that the Moon, at its origin, was completely melted in magma, although there is no definitive evidence to support that the Earth was melted to a similar degree[28]. Yet, again, the isotopic analysis shows that both bodies have a similar age, around 4,500 million years. Why only the smaller planetary body, which can cool more quickly, shows evidence that was completely melted, but the largest body was perhaps not fully melted in a similar way?

More mysterious still is the fact that the analysis of the magnetic field intensity of the Moon and its variation with time, which is reflected on some characteristics of the lunar rocks, indicates that the iron core of the Moon was formed millions of years after the Earth's[29]. In addition, other data indicate that the mantles of Earth and Moon are slightly younger than the iron core. How can we explain this?

Other oddities, this time not only proper to the Moon, but to the Earth-Moon system, come from its physical features, differently to the case of the chemical features explored above. We must first consider the size of the Moon. The Moon is very large relative to the Earth, which, by the way, allows tides to be produced. To get an idea of how big the Moon is, it suffices to say that our satellite is the fifth largest in size of the entire Solar System. The largest, Ganymede, orbits around Jupiter; the second, Titan, around Saturn; the third, Callisto, and fourth, Io, orbit, again, around Jupiter. Even the giant planets Uranus and Neptune do not have satellites competing in size with the Moon. In addition, the other planets comparable in size to the Earth either do not have satellites, like Mercury and Venus (the only planets in the Solar System lacking satellites), or have them of a size too small to exert any significant gravitational influence on their planet, like Mars, which has two very small satellites the size of asteroids, called Phobos and Deimos, unable to exert significant tidal effects[30]. How come the Earth has a satellite so large relative to its size?

Another oddity of the Earth-Moon system is its high angular momentum. Angular momentum is a physical property related to the rotational energy of the physical bodies. The rotational energy of the Earth-Moon system is very high compared to that of other planets and satellites. If we know something about the universe is that energy is not free and must come from somewhere. Where does the energy that has provided such high angular momentum to the Earth and Moon come from?

Finally, another oddity of the Earth-Moon system is that the Moon's orbit is tilted about 5° relative to the rotational axis of the Earth. This does not happen in the case of other planets and satellites[31]. The Moon is closer to the ecliptic, the plane on which the planets orbit the Sun, than to the Earth's equatorial plane. How has the Moon acquired the tilt of its orbit? Again, this requires an input of energy, which had to come from somewhere.

The best hypothesis

Which one of the hypotheses on the origin of the Moon mentioned above best explains all its features and oddities: size, chemical composition, tilted orbit, and angular momentum? Let's analyze the advantages and disadvantages of each of the scenarios discussed above, beginning with the capture hypothesis proposed by Descartes in the first place.

The capture hypothesis suffers from two major drawbacks to explain the origin of the Moon. The first weakness is that, as explained, isotopic analysis of Moon rocks indicates that the Moon had to be formed at the same region of the Solar System as the Earth formed. The second problem is that computer simulations and calculations of the orbit and energy that the wandering Moon should have possessed to allow its capture by the Earth's gravity and, paradoxically, thus stop being so lunatic, indicate that this capture could only occur if a number of very extraordinary circumstances occurred. If the Moon had approached the Earth at a relative speed of the order of miles per second, as often happens with other astronomical objects (comets or asteroids) passing along the vicinity of Earth, it would have been impossible to Earth's gravity to capture it, considering besides that, in that far-away epoch, Earth had only 2/3 of the current mass because it was so young that it was still growing. The only way in which the Moon could have been captured by Earth's gravity is if it had approached the Earth at a very low relative speed. But this is very, very, unlikely. So, if we consider the two problems together, i.e., the identical isotopic composition of Earth and Moon, and the difficulty of the dynamics of capturing by gravitational pull a body the size of the Moon, we have to abandon this hypothesis as a plausible one to explain why the Moon orbits the Earth. It is, in fact, what scientists studying the origin of the Moon have done. Lets then discard the hypothesis of Descartes as a valid one, and focus our attention elsewhere.

What about the fission hypothesis? Let recall that this hypothesis, proposed by Darwin's son, claims that the Moon detached from the Earth due to its initial higher spinning speed, helped by the Sun's gravitational pull. This hypothesis has the advantage of explaining why the Moon has a very small iron core, because if it came out from the Earth's mantle, when most of the iron had fallen to its center, it is expected the iron content of the Moon to be much smaller than that of the Earth. Unfortunately, remember that the volcanic rocks of the Moon lack water and that the Moon, overall, has a smaller content of light chemical elements in the upper mantle than the Earth. This should not be so for a satellite detached from its parent planet.

Another major difficulty faced by this hypothesis is that modern calculations show that to allow the uprooting of the Moon from the Earth, our planet must have rotated a minimum of once every two and a half hours[32]. This has been proven to never happen, not even when Earth was only a protoplanet. But, in addition, we need to remember that the Moon's orbit is tilted and not located on the Earth's equatorial plane, which should almost certainly not be the case if this hypothesis were true. It is possible yet that after splitting off from Earth and settling in an equatorial orbit, the Moon had slowly migrated to reaching its current orbit, but to allow this to happen it is necessary to invoke complex and unlikely mechanisms. In conclusion, this hypothesis, although it explains some features of the Moon, such as its small iron core, it does not explain others and poses so many problems (more still than those mentioned here) that it has been also abandoned.

Let's now consider the hypothesis of the co-formation of the Earth and the Moon. This hypothesis, in light of what we know, might explain their similar chemical composition, but it raises other problems. To begin with, it cannot explain why the Moon's orbit is tilted about five degrees to the plane of the Earth's orbit around the Sun (called the ecliptic plane), and also why the Moon does not rotate on the Earth's equatorial plane. If the Moon formed near the Earth it should rotate on this plane, and if it formed far away, it should rotate on the ecliptic plane. This hypothesis cannot explain either the high angular momentum of the Earth-Moon system, and cannot explain why the surface of the Moon was once completely molten magma. The formation of planets and satellites of the size of Earth and Moon by coalescence of scattered matter do not provide enough energy to allow the surface temperature reaching that point. It is also impossible for this hypothesis to explain the different matter densities of the Earth and the Moon. If both were formed from the same cloud of matter, the density of the two should not be as different as it is. Finally, another problem posed by this hypothesis is that it does not adequately explain the large size of the Moon relative to Earth, since being formed at the same time, as has happened with other satellites and planets, the Moon should be much smaller. In conclusion, again this hypothesis allows explaining some things, but it is incompatible with many of the characteristics of the Earth-Moon system. Therefore, the hypothesis of co-formation must also be abandoned. This leaves us only with the hypothesis of the collision.

In his adventure *The Sign of the Four*, Sherlock Holmes said to his beloved Dr. Watson: "How often have I said to you that when you have eliminated the impossible, whatever remains, however improbable, must be the truth?" This is exactly what happens with the hypothesis of the collision. The other three hypotheses have been discarded as impossible, there is no other alternative hypothesis that we can consider, therefore, what remains, however improbable, must be the truth.

And the truth is that it must be true. Besides, the collision hypothesis offers many advantages, and practically solves all problems, despite it may appear as very unlikely that the Moon originated from one, necessarily, cataclysmic collision that relegates the collision that led to the extinction of the dinosaurs to a simple anecdote of no importance in the history of the planet.

The main advantage, in my view, of the collision hypothesis is that it leaves open all possibilities to consider the conditions under which the collision had to happen to originate the Moon. That is, it allows us to play with variables such as size of the object that collided with the Earth, its speed, angle of collision, temperature, etc. These variables can be selected in various computer-simulated models to identify only those that are compatible with the current characteristics of the Earth-Moon system. Only in the event that the collision models are not consistent with observations we should abandon this hypothesis, but I tell you right now that this has not happened, and that this hypothesis is the one currently accepted by the scientific community to explain the origin of the Moon.

The reason is that the collision hypothesis can explain virtually all the observations. It can explain why the Moon does not orbit on the ecliptic plane, or on the Earth's equatorial plane, as the collision originated a rotational system unlike any other in the Solar System. It can explain why the lunar volcanic rocks lack water, as it was lost due to the high temperatures that such a great impact produced. For the same reason, it can also explain why the surface of the Moon was completely melted into magma, and suggests that the

Earth's was equally so, although we possess no definitive evidence to support this claim. It can also explain the identical oxygen isotope ratio found in the Earth and Moon, and why the Moon contains a smaller proportion of light chemical elements than the Earth. It also explains the high angular momentum, or rotational energy, of the Earth and Moon, which was acquired at the expense of the energy of the collision. Finally, if I do not forget anything, it also explains the small iron core of the Moon and why it formed after the Earth's, because the Moon came from material located at the Earth's mantle, and the mantle of the planetoid that collided with it, when iron had already migrated to their cores. It is true that the core of the planetoid that collided with the Earth probably was also rich in iron, but what the computer–simulated collision models suggests is that its iron core merged with that of the Earth, so that iron did not feed in large quantities the material placed in orbit after the collision, which later would coalesce in a single planetary body: the Moon.

The great collision

What do modern studies and calculations tell us about the way this collision happened?

Several years of study by many scientists and massive computer calculations were necessary to develop a model of the collision that gave birth to the Moon. According to these calculations, it is estimated that a planetoid of a size similar to that of planet Mars also formed along with the other planets in the solar system, about 4,500 million years ago, take or leave a minute. This planet has been appropriately named

Theia, mother of the Greek goddess Selene, goddess of the Moon. Theia is, therefore, the Greek grandmother of the Moon. Interestingly, this protoplanet formed in the same orbit as that already occupied by the proto-Earth; however, the gravity of this latter and larger body did not absorb nor capture the material that formed it. Is this possible?

Well, yes it is. That's because each orbit involves at least two bodies gravitationally attracting each other. For example, the Earth's orbit around the Sun is not only the result of the attraction of the Sun exerts on Earth, but also the result of the attraction the Earth exerts on the Sun, although this is much lower. As a consequence in each orbit of each planet there are points where the gravitational attractions of Sun and Earth cancel each other and in which another body can settle stably (at least for a while) in relation to the other two bodies. These points are called the Lagrange points[33], named after the French mathematician who discovered its existence. In these orbital points, a body with a mass below a certain threshold experiences no gravitational force as far as it follows the orbit of the major planet. There are five Lagrange points for each orbit, and two of them are in the same orbital path followed by the main planet, forming an angle of 60° to the direction of the Sun. It is believed that Theia formed in one of these points[34].

Why did Theia leave that point of orbital stability? As I hinted earlier, a body can continue at a Lagrange point while not exceeding a given mass limit. If exceeded, the gravitational pull is too strong; it interacts with other bodies in the orbital system and leaves the Lagrange point. Thus, one reason that could induce Theia to abandon its orbit is that, during that epoch of planet formation in the Solar System, it grew above this mass limit, causing a stronger interaction with Earth that forced it to leave the Lagrange point and take a collision course with our planet. Another possibility to consider is that Theia might have collided with another massive body and thus be forced to leave its orbit. In any case, the formation of Theia on the Earth's orbit means that the material that formed it was almost identical to that which coalesced to form the Earth. This explains the identical isotopic composition between the Moon and Earth, of which we spoke earlier.

In what manner did the collision between Earth and Theia occur to end up in the formation of the Moon? Computer models developed by Dr. Robin Canup enlighten us about this issue[35]. These models[36] show that for the collision to give rise to a satellite like the Moon, it had to happen not once, but twice!

In the first collision, Theia hit the Earth at an oblique angle. As a consequence, the iron and nickel core of Theia fell down to the center of the Earth and merged with the core of it. What was left of Theia, mainly the mantle of the planet, slowed down in its progression because of this first collision, fell down again to Earth in a second and final collision, which caused the complete destruction of Theia. These two collisions occurred in a single day and caused the release to space, both from Earth and Theia, of matter that later coalesced to originate the Moon, which formed at a distance less than fifteen times the current distance, i.e., only at about 22,000 km (around 13,500 miles) from Earth. The impact significantly increased the speed of the Earth's rotation, i.e. its angular momentum. This explains the high angular momentum of the Earth and Moon system. The day, in those days, lasted just four of our current hours, but day length increased and continues to grow slowly today because, due to the effect of the tides, the Earth is still transferring angular momentum to the Moon, which still continues to move away from us, at the same time the Earth increases its spin period[37].

Let us stay a bit more in the aftermath of the collision. The energy of this titanic collision was so huge that, despite we lack experimental evidence to confirm it, as we said, it is reckoned that the temperature increased to a level high enough to melt all the rocks on the surface of the Earth and the matter that latter would become the Moon. Thus, it is estimated that almost the entire Earth's surface was transformed into an ocean of magma that took a million years to cool down. Until this ocean cooled down to a temperature low enough to allow water to remain in the liquid state, it was not possible for life to appear on Earth. Thus, the moment when Earth and Theia collided is, in fact, the zero time point from which life could develop on our planet.

An important consequence of the collision was the likely loss of water released into space. We do not know how much water our planet lost, but some scientists estimate that had it not lost any, probably the continents would not exist and our planet Earth should be called, more so than now, planet Water.

The loss of water due to the collision could have been decisive, therefore, for the existence of continents over the Earth's surface. In particular, considering that about 4,100 to

3,900 million years ago, Earth and Moon suffered an intense bombardment of celestial bodies, which formed the majority of craters that can be observed today on the surface of the Moon. This is called the Late Heavy Bombardment[38]. It is unclear what kinds of bodies were involved in the bombing, especially if they were mainly asteroids or mainly comets. This is important because whereas comets are composed mainly of ice, asteroids are either rocky or metallic. Data published in July 2009 suggest[39] that the bombardment was caused mainly by comets, which could have brought no less than 3,400 tons of comet material per square meter of the Earth's surface (around 317 tons per square foot), that is, enough water to form the current oceans and seas on our planet. If this is true, this means that had water not been lost in the collision with Theia, the Earth would probably be completely covered with it and continents would not exist. The collision that caused the Moon, therefore, could have played a critical role in shaping the conditions of Earth, enabling the presence of large continents on its surface rather than perhaps a few scattered islands, so that intelligence capable of creating a technological civilization might one day develop.

Besides, the role of the continents is not limited to providing a platform on which life beings could develop a level of intelligence leading, perhaps, to the birth of a technological civilization. Their role has also been critical to "fertilize" the oceans with the metals and minerals necessary for the development of more complex life forms. One such metal is molybdenum, necessary for the activity of enzymes that catalyze the fixation of atmospheric nitrogen by some microorganisms and plants. Nitrogen fixation is essential for the synthesis of amino acids, which are the components of all proteins. In the absence of contact with oxygen, the oxidation of molybdenum and its dissolution in water is impossible. Without continents, the minerals of the Earth crust would have been under water, and oxidation of molybdenum would have happen much more slowly. This would have made very difficult its transfer to primitive living organisms and could have slowed down the emergence of more complex forms of life, which need the fixation of atmospheric nitrogen to synthesize proteins in sufficient amount to allow their development[40].

Another effect of the birth of the Moon and its presence around the Earth is that it could have protected it from collisions with comets or asteroids, subsequent to those occurred at the Late Heavy Bombardment, that might have produced mass extinctions, or even the entire annihilation of life. Any mass extinction that the Moon could have prevented might have changed the course of evolution on the planet, making impossible perhaps the appearance of a species like ours or, at least, delaying its onset and, therefore, the emergence of a technologically advanced civilization. Probably some asteroids and comets, which might have collided with Earth, met in their way the Moon instead. Collisions were more frequent when the Solar System was young, when life had just emerged on our planet and was probably more fragile and sensitive to the effects of major hits that could have caused its complete extinction. However, when the Solar System, the Earth, and the Moon were young, because of the way the latter formed, the Moon was, as stated above, much closer to Earth. This closeness to Earth significantly increased its capacity to protect it from collisions with other astronomical objects.

Another important consequence of the origin of the Moon has been, and still is, that the Moon's gravitational effect has decreased the fluctuation along time of the angle formed by the Earth's spinning axis with its plane of rotation around the Sun, i.e. the ecliptic plane. As we know, the axial tilt of Earth's spin axis to the ecliptic plane is 23.44°. Due to this, the Earth's climate undergoes the phenomenon of the seasons. As the planet continues its path around the Sun, the tilt of its axis causes the north or south hemispheres of the Earth to be facing the Sun (it occurs in spring and summer), or facing away from it (autumn and winter). For this reason, when summer begins in the northern hemisphere (axis tilted 23.44° toward the Sun), it is winter in the southern hemisphere (axis tilted 23.44° away from the Sun).

The axial tilt of the Earth respect to the ecliptic has a great influence on the intensity of the seasons. If the Earth's axis were tilted a steeper angle, say 45°, summers would be much hotter and winters, much colder. With a 45° tilt, in summer, on all regions north of this latitude the sun would never set down for a period that will be shorter or longer depending on the actual latitude. Thus, during a period of the year the sun would warm the Earth's surface of those latitudes literally the entire day, and the temperature would consequently increase. In fact, under these conditions, summer temperatures would be much higher, reaching up to 50°C, 60°C (122 - 140 °F) or more at noon, and they would not drop much during the short nights. By contrast, with an axis tilted 45°, at the beginning of winter on the same regions the Sun would not raise and we would have a perpetual night. Lacking solar heating, temperatures would fall dramatically and winters would be much colder and harder. Thus, the axial tilt of the Earth affects how much hot or cold the summers and winters are, and the extremes of climate variation between these two seasons.

On the contrary, if the Earth's spin axis were not tilted, it would be always spring (or autumn) everywhere. The sun would be over or under the horizon exactly twelve hours in all latitudes throughout the year, and temperatures would not fluctuate as they do now with the seasons, because there would be no seasons.

Thus, a different axial tilt would change the Earth's climate more or less sharply, depending on the difference to the current tilt angle. In fact, these climatic variations occur periodically, because, in reality, the axial tilt of the Earth is not constant, i.e. it is not always 23.44°, but varies from 22.1° to 24.5° cyclically over a period of 41,000 years[41].

This variation has exerted a major impact on the ice ages suffered by the Earth, and perhaps it will influence future glacial periods. These occur more easily if summers are not too hot and fail to melt completely the ice and snow fallen during the winter. Snow and ice act as reflectors of the Sun's energy and prevent it from being absorbed by the planet. Snow and ice increase the so-called albedo of the Earth, i.e. the proportion of sunlight reflected out by the planet[42].

If one Earth hemisphere does not get hot enough in summer, and it still keeps ice and snow, when winter comes again more snow and ice will accumulate, and they will be still more difficult to melt the following summer. Gradually the ice builds up and an ice age can start.

The lesser the axial tilt of the Earth, the colder the summers will be, as the area of the planet tilted toward the Sun during the summer will be less. This means that when the Earth's axis is tilted the least with respect to the ecliptic (22,1° as discussed earlier) glaciations will become more likely. However, when the axial tilt is greater (up to 24.5°), glacial periods will be less likely.

Currently the Earth is in the part of the cycle in which the tilt angle increases. That is, we are heading slowly to hotter summers and colder winters (not taking into account other factors that may affect the climate, such as global warming caused by the emission to the atmosphere of greenhouse gases). Therefore, we are getting away from the danger of glaciations.

What would happen with the axial tilt of the Earth if it lacked its satellite, the Moon? What scientists believe it would happen or it would have already happened, is that the Earth's axial tilt would have varied much more greatly throughout time. This variation would have translated into more severe climate changes. The Earth's climate could have been extreme, and if that might have not affected the rise of life in the oceans of our planet, it could have affected the conquest of the continents by live organisms. The effects of climate change are most evident on the Earth's surface, not in the oceans. This is so because water has a high capacity to absorb or release heat without varying its temperature too much[43]. At the North Pole, whereas the surface is obviously the polar ice cap, the bottom of the ocean is liquid, and the temperature is, obviously, above the water's freezing point. The polar winter affects far more dramatically the Earth's surface than the seabed, and, in summer, again water temperature does not raise the same way as that of the air.

The above means that if the Earth's axial tilt fluctuated in a more important way, the organisms that would have dared to venture out of the water and onto the continents would have met more drastic climate variations. If that might not have prevented completely the colonization of land by life, it probably would have delayed it or difficult it, which in turn could have changed the course of life evolution on the Earth's surface. For this reason also, perhaps without our Moon we would not be here now.

Finally, a factor already discussed that changed the conditions under which life evolved on Earth is the phenomenon of the tides, which in the absence of the Moon, being caused only by the Sun, would have been much smaller. We have already discussed above the importance of tides to create ecological niches that facilitate the colonization of the land by life. Furthermore, due to the short distance from the Earth at which the Moon formed, and to the rapid rotation of the Earth, the Moon-caused tides were initially more intense and of shorter periods than they currently are. This might be important for two reasons. First, the increased intensity of tides results in a greater amount of land inundated by them, which possibly extended the ecological niche where life could adapt to the land environment. This reason, however, is not the most important because this great ecological niche occurred early in Earth history when there were probably not pluricellular animals capable of tackling the conquest of land. The second reason, more interesting, is that, being hundreds of millions of years ago of a shorter period than they are today, the tides made it possible for the organisms to survive with a greater likelihood the short period of low tide, when they were in the absence of water. That is, when the Earth, thanks to the angular momentum acquired in the collision with Theia, rotated over its axis once every four hours, the tides rose and fell every hour, and did so with great intensity due to the proximity of the Moon. As the Earth spun more slowly, and the Moon moved away, the tides fell in intensity and lengthened their period. However, when sea life started to colonize the land, tides were always more intense and of a shorter period than today. In addition, it has even been speculated that just after the formation of the Moon, the rapid wetting and drying cycles that the tides caused were important for the replication of the first self-replicating molecules that gave rise to life[44]. If so, certainly the origin of the Moon would have exerted a major influence for life on Earth.

Moreover, the short period of our tides is only possible because of the increased spinning speed of the Earth caused by the collision with Theia. Had there been no such collision, and, therefore, had not the Moon been originated, Earth would spin much more slowly around itself. This slow rotation would have caused that the tidal period caused only by the Sun, even billions of years ago, when life arose on our planet, be much longer than it is even today. Consequently, the ecological niche in which plants had to adapt to life on land would have been much more hostile, as the period the tides would had left the coastal areas uncovered of water would had been, perhaps, too long to allow an easy survival of the organisms that could have ventured into these areas. The rapid rise and fall of tides not only allowed by the presence of the Moon, but also by the rapid spinning of the Earth as a result of the collision which formed the Moon, facilitated the adaptation of marine plants to the land, because it did not leave them too long in the absence of water. The rapid rise and fall of tides would not exist had not happened the collision with Theia that created the Moon.

Another important effect caused by the collision with Theia, related also to the tides, is that the rapid rotation of the Earth on its axis allows the rapid transition between day and night. This allows daytime temperature not to raise much during the day, or fall down much at night. If the rotation of the Earth around itself were as slow as that of Venus, which rotates around itself in 243 Earth days, the temperature at "high noon" and early "afternoon" on the Earth would be of hundreds of degrees, but the "midnight" temperature, less than a hundred degrees below zero. For comparison, we can look at the temperatures during the Moon's days and nights, which last about 14 Earth days each. During the lunar day the temperature can reach 123°C (253,4°F), and 181°C below zero by night (-293,8°F)[45], close to the liquefaction temperature of oxygen. That is, the difference in temperature between dawn and noon/afternoon on the Moon is about 280°C (536°F). In these fluctuating temperature conditions the evolution of life on Earth would had been more difficult, and much more difficult, if not impossible, the colonization of land by it. Our civilization, almost certainly, could not have developed if the

Earth's rotation were significantly slower than it is today, thanks to the collision that accelerated it.

These effects exerted by the Moon also limit the number of planets with no moon in which civilizations could develop, if this were possible on these planets. To possess strong and fast tides, a planet lacking any moon should revolve at a short distance around its star, because as tides are caused by gravity, tidal intensity decreases very quickly in a direct relation to the cube of the distance to the central star. That is, doubling the distance from the star decreases tidal intensity eight times. This implies that the central star cannot be very hot, and thus not very bright, or else the planet will have no liquid water on its surface, being too close to it. But beyond this limitation, we have an even more important one: due to the effect of the tides, the gravity of the star tends to slow down the rotation of the planet around itself and, little by little, causes that the planet ends up facing the star, in a phenomenon, called tidal lock, similar to which it has happened to our Moon, which ended up showing the same side to Earth just a few thousand years after their formation[46]. This implies that the putative planet will not have enough time for life to evolve on it, let alone intelligent life, before the effect of tides sets it in a fixed rotation around itself, always facing its star. This, in effect, is what happens with the first extrasolar Earth-like rocky planet discovered by the COROT mission[47]. This planet revolves at only 2.5 million km (1.55 million miles) from its star (in comparison, the Earth rotates on an average 93.2 million miles from the Sun), and it is constantly facing it. The temperature on the side directed toward the star raises up to 3,000°C (5,432°F), while the other side only reaches temperatures of -220°C (-364°F). Clearly, under these conditions, life is impossible.

Even if a planet revolved farther away from a dim star, but close enough to have liquid water and significant tides, this would leave a period of, at most, only a few hundred million years before the tides ceased to exist in this hypothetical planet, and its days and nights became extremely long, resulting in wide temperature fluctuations[48]. As noted, this period is probably insufficient to allow the evolution of multicellular plants that can later colonize the continents and allow animal life on them, the only kind of life capable of developing intelligence. Probably, that moonless planet orbiting a "dimly lit" star would constantly face it and, therefore, it would lack tides when (if) marine plants were sufficiently developed to dare trying colonizing the land, if land there is. In these conditions, they would not receive the help of this evolutionary space-time generated on our planet by the Moon. Besides, this would only happen in the event that the little light emitted by the dim star allowed photosynthesis, in the lack of which multicellular life would probably be impossible. In addition, even if photosynthesis were possible, the extreme temperature difference between the planet's hemisphere directed toward the star and the opposite hemisphere would only allow a slight stretch between the two in which the temperature might be adequate to allow liquid water. However, in these conditions it becomes difficult to imagine that life in this strip would evolve to reach intelligence, much less to develop a civilization.

All right, you would think, my informed reader: those considerations are not relevant in the case of planets revolving around brighter and therefore hotter stars than the Sun. The habitable zone of these stars is much wider than the Sun's, extending hundreds of millions of miles[49]. Planets located into this habitable zone could develop life, if they had the right conditions to allow liquid water on their surface. You are right. However, it would be very difficult, even with satellites like the Moon exerting a gravitational effect and causing significant tides, that a civilization developed on these planets if an amount of time similar to the one required to develop our own civilization were also needed (not considering that without a moon this will probably would take even more time). The reason for this is simply the short longevity of the brightest stars. The larger and hotter, the shorter the longevity of stars,[50] and this may prevent enough time to develop sufficiently advanced beings and, therefore, the development of a civilization. For this reason, some astrophysicists and astrobiologists claim that civilizations can only exist around Sun-like stars (only accounting for 5% to 10% of all stars), which have a high longevity[51]. Besides, these stars do not deliver high doses ultraviolet radiation or X rays, which could affect the stability of the complex molecules of life, especially when life ventures to colonize the land. Under these conditions, the influence of a moon around the planets can be decisive.

Thus, the collision that created the Moon has enabled a rapid transition between day and night, avoiding extreme temperatures on the surface of the Earth and allowing short period tides, which have facilitated the adaptation of plants and animals to the land. It has also allowed quick tides, of considerable intensity, in the period in which multicellular plants came into existence, about three billion years after the origin of life on our planet, a time at which, had not happened the collision with Theia that created the Moon, Earth would probably offer the same side to the Sun, or otherwise it would not possess a rotational period short enough not to cause very long tidal periods, unfavorable to accelerate the evolution of plants toward the conquest of the mainland and thus, towards the development of intelligence and hands, allowing civilization to develop. In addition, the rapid rotation of the Earth around itself, made possible by the collision with Theia, causes the temperature difference between day and night to be much less than it would be if the Earth rotated more slowly. This greater homogeneity of the Earth's temperature is very favorable to the development and evolution of life on its entire surface. Definitely, without our Moon our chances of being here now talking about it would have been much lower.

Are collisions frequent?

Thus the birth of the moon through a huge collision with the protoplanet Theia had profound implications for Earth history. Now that we have determined that the origin of the Moon appears to be a huge impact, we can try to estimate the answer to the question that led us to this tour by the selenology: what is the frequency of moon-like satellites around planets like the Earth that could exist in the Universe? I hope that, by now, you accept the idea that, according to this frequency, the frequency of technological civilizations joining us on other planets would be higher or lower. For this reason, this is a relevant question.

Scientists know by now that the frequency of asteroid, meteorite or comet collisions with the Earth is inversely proportional to their size[52]. That is, the smaller the asteroid, the greater the frequency of collisions with them. Millions of meteors the size of small pebbles collide with the Earth's atmosphere every day. Thanks to observations with satellites and other advanced media, it is now estimated that an asteroid about 5 to 10 meters in diameter hits the Earth once per year, on average[53]. These collisions release as much energy as the atomic bomb that destroyed Hiroshima. However, these collisions pass generally unnoticed because the explosions they cause occur in the atmosphere, when the asteroid is vaporized and destroyed tens of kilometers above the Earth's surface, more frequently over the ocean or over uninhabited continental regions, which still constitute the bulk of the planet's surface even though we are already close to seven billion human beings on it. For example, in the summer of 2002, U.S. satellites detected an explosion on the Mediterranean similar to that caused by an atomic bomb[54]. Fortunately, scientists were able to determine that its origin was the collision with an asteroid. More recently, on October 8, 2009, again the detection network of nuclear explosions detected the breakup of an asteroid about 10 meters in diameter on the Indonesia's sky[55].

Collisions with larger bodies, capable of penetrating the atmosphere and reach the Earth's surface, are less frequent. It is estimated that collisions with objects around 50 meters in diameter occur at a frequency of once every one thousand years. It is believed that one of such collisions occurred in a region near the Tunguska River, in Russia, on June 30th 1908[56].

Asteroids with a diameter of 1 km (0.62 miles) impact the Earth once every 500,000 years, on average[57]. Collisions with greater asteroids of about 5 km (3.1 miles) in diameter might occur every ten million years. The frequency of collisions with larger asteroids is even smaller. The last known collision with an asteroid of about 10 km (6.2 miles) in diameter occurred, fortunately, 65 million years ago, and caused the extinction of the Cretaceous-Tertiary period, including the ever-famous dinosaurs[58]. By the way, I said fortunately because had this collision not happened, we probably would not be here to experience the immense pleasure of reading these magnificent lines (please, note the ironic tone).

Considering the above data, one is tempted to reach the conclusion that if the bigger the more unlikely, then collisions between planet-sized objects must be extremely improbable, and perhaps only happened here and there in scattered places of the universe, in a very, very, small number of solar systems. However, we have evidence of other major collisions in our Solar System, in addition to that which formed the Moon. As an excellent example, our own Moon keeps the considerable sign of a huge collision: the crater located at the lunar south pole (called the Aitken depression[59]), of about 2,500 km (1,553 miles) in diameter and the second largest crater in the Solar System. This crater does not appear to be the result of a high angle collision, but the result of a low-angle impact, i.e. more direct than that occurred between the Earth and Theia. In any case, this collision did not leave behind a satellite of the Moon, nor of the Earth.

The largest crater in the Solar System was confirmed in 2008[60] and is located at the northern hemisphere of Mars. This enormous crater is about 10,600 km (6,600 miles) long and 8,500 km (5,300 miles) wide. It is estimated to have been caused by a major collision occurring nearly four billion years ago. Mars also possesses the third largest crater in the Solar System, known as the Hellas Planitia[61], of about 2,200 km (1,370 miles) in diameter. The NASA Messenger spacecraft also discovered, in 2008, a crater 715 km (450 miles) in diameter on the surface of Mercury[62].

The presence of these large craters suggests that major collisions were frequent in the early Solar System. In fact, modern theories about Solar System formation estimate that initially, in the inner region of the Solar System between Mercury and Mars, 50 to 100-protoplanets of sizes ranging from the Moon's to Mars's, were formed[63]. The collision and fusion of these protoplanets was what eventually led to the formation of the four inner planets of the Solar System we all know, which, in the case of Venus and Earth, are of sizes considerably larger than those estimated for the initial protoplanets.

This suggests that the collisions leaving the large craters on Mars, Mercury and the Moon happened at the end of the planet formation period. We have already mentioned that the collision with Theia happened when the formation of Earth could be considered complete, i.e., when the period of collision and fusion between the protoplanets initially formed had finished. Posterior mergers or collisions, including the collision with Theia, possibly eliminated the remains and effects of previous collisions. That is, although collisions may be initially frequent, only those that occurred towards the end of the formation period of the Solar System left a lasting effect that could not be disturbed by subsequent collisions with other protoplanets.

An example of this might well be what it is believed happened with planet Venus, which although similar to Earth in size is devoid of any satellite. The reason why Venus has no satellite could be because at the end of their formation it suffered not one, but two collisions with two different protoplanets[64], (and not two with the same protoplanet, as it happened with the Earth). According to this hypothesis, based on computer simulations that take into account the current parameters of rotation and angular momentum of the planet, Venus collided with a large protoplanet, which might have caused a moon similar to that of Earth and also accelerated the spinning around its axis, as it happened with the Earth. Unfortunately, only 10 million years later, Venus suffered another collision in the opposite direction, which slowed its spinning speed (now the "day" of Venus lasts 243 Earth days) and, due to gravitational effects, caused that the satellite generated in the first collision and the debris generated during the second one, ended up falling on the planet, which thus lost its moon. Anyway, with or without a moon, Venus is too close to the Sun so that life could develop on it.

In any case, if these two collisions really occurred, what happened to Venus, and the large craters seen on other rocky planets in our Solar System, suggests that major collisions between protoplanets can be frequent, but for them to give rise to stable satellites, as the Moon, they must happen in a closed range of conditions and toward the end of the planet formation period. Considering what we see in our Solar System, the most likely outcome of planet formation is the generation of planets lacking satellites. The only exception seems to be the satellite of Pluto, Charon, whose origin seems to be due to a collision between Pluto and another body of similar size. In any case, the generation of a planet with a large satellite like the Moon, and located in the habitable zone of the star, seems unlikely.

It is possible (and what follows is my own independent hypothesis, which, however, I have also seen postulated in some publications) that the last major collisions in the period of planet formation are the result of planetoid formation at the orbit's Lagrange points, as explained above. At such Lagrange points, at one of which, in the Earth's orbit, is estimated that Theia formed, the gravitational attraction exerted by other bodies is canceled out, allowing mass accretion and the formation of minor planets. However, to keep these planetoids in the same orbital Lagrange point, the masses of the other orbiting bodies and, most importantly, the mass of the body located at the Lagrange point, should not change. If these masses change, so do the gravitational forces, resulting in the appearance of instabilities inducing that the bodies located at these points leave the orbit.

At the origin of the Solar System, when planetoids were in a growing phase by accretion and accumulation of matter from the protoplanetary cloud surrounding the young Sun, the bodies located at the Lagrange points were in a situation of orbital instability. This suggests that bodies formed in those points would eventually abandon them, and since they were in the same orbit of a major protoplanet, they could maybe get into a collision course with it, although they could also enter a collision course with the Sun, that would absorb them, or be lost in space (planets lost in space without any fixed orbit have been already detected near some young stars[65]). In the case of Theia, we were fortunate that it collided with Earth and the Moon was formed, which might not be the most likely outcome given the immense interplanetary distances when compared with the sizes of the planets and the stronger gravitational attraction exerted by the Sun as compared to that exerted by Earth.

So, I believe it reasonable to assume that despite big collisions might be significantly more likely at the origin of the Solar System than they are today, and that they may be also more likely at the origin and formation of other planetary systems, very few among them will be happening with the conditions of angle and speed required to form Moon-like satellites around planets similar to Earth, orbiting around a star similar to the Sun, providing the necessary conditions for the development of life.

Persisting on this idea, what I believe is true is that despite the fact that collisions between protoplanets may be relatively common, so that each planet might have suffered two, three or even four collisions before getting into a stable orbit, the collisions that might generate a satellite of sufficient size to gravitationally affect the central planet must be the last one to happen and are very rare, perhaps unique. We have here the apparent paradox that probable events, such as collisions between protoplanets during the formation of planetary systems, do not lead often to double planetary systems similar to the Earth and the Moon. In fact, none has been discovered outside the Solar System when I write these lines, although it is possible this may be also due to the current difficulty to detect them. As I said, this may seem paradoxical, but actually it is not. Every day thousands of children are born; billions of people have lived on Earth, but only one of them was born with the necessary conditions to become Einstein, and only another one with the conditions to become Picasso. Billions of collisions between protoplanets may also have happened in the Universe, but only a few might have occurred with the necessary conditions to generate large moons around planets that also possess the conditions necessary to sustain life.

For that reason, I would like to examine the subject of collisions between protoplanets a little further. What would have happened if the collision with Theia and the Proto-Earth had happened with a different angle or with a different speed? Clearly, if any collision parameter had varied (including the mass of the colliding bodies, their speed, their chemical composition, and the impact angle) the result would have been different. For example, if the bodies had collided at a higher speed perhaps the matter ejected into space could not have coalesced again to form the Moon, as matter can coalesce again only if it is thrown away to space with a speed lower than the escape velocity of the gravitational system. If as a result of a stronger collision, matter would have been thrown out with a speed exceeding the escape velocity, gravity could not have held it and now it would be lost in outer space.

Similarly, a different collision angle or speed could have generated, for example, a smaller Earth and a greater Moon. This, in principle, may have not changed the conditions for the development of life and civilization on Earth. Initially, tides would have been even more intense, further promoting the conquest of mainland by living organisms. In addition, being the Moon greater, it may have possessed enough gravity to retain liquid water and life could also have developed on its surface. However, a smaller Earth and a greater Moon would have significantly shortened the period in which both bodies would have reached what is called the tidal lock, a situation in which both would spun around their axis with the same period they would revolve around each other. This is what has happened in the case of the dwarf planet Pluto and its moon, Charon. In addition to making the tides disappear soon, perhaps too soon to help conquer the land by living organisms, the tidal lock would have lengthened also the duration of days and nights, causing wide variations in temperature that would have made more difficult the conquest of land by life.

Another possibility is that a different collision angle or speed could have generated a greater Earth and a very small Moon, perhaps too small to exert a significant gravitational effect on Earth. A small Moon would not produce significant tides, not would stabilize significantly the Earth's rotational axis, so its accelerating effect on the conquest of land by life would have been very small, or inexistent.

In addition to the fact that various collision conditions could have generated a different Moon, or not generated any, recent studies on the planetary systems that are being discovered around other stars indicate that some of the planetary orbits are very eccentric, i.e. the orbits have the form of stretched ellipses. The planet passes, therefore, very close to the star at a time of its orbital path and is far away from it at another time[66]. These eccentric orbits appear to occur as a result of major collisions that happened on the final moments of protoplanetary formation. Thus, a consequence of the collisions that occurred during the late period of the Solar System formation could have been a planet Earth and its satellite with a much more eccentric orbit than the current one. In this case, temperature would have varied so much throughout the year that it would have made more difficult the conquest of land by living beings and, therefore, it would have hindered the development of intelligence leading to a technological civilization.

Thus, it seems clear that the occurrence of collisions between protoplanets during the formation of planetary systems around other stars cannot happen in a large variety of conditions so that they give rise to a double planetary system, with a major planet and a smaller one, but large enough to gravitationally affect the other without slowing down significantly its rotational speed. We know that the largest Solar System's rocky planets are Venus and Earth, so perhaps the size of rocky planets is also similar in other planetary systems around Sun-like stars. This is what seems to indicate the discovery of a new solar system at two thousand light-years from earth[67]. This may limit the size of the colliding bodies, causing bodies like Earth and Venus being the biggest with which others can collide. In any case, few collisions are likely to occur with the appropriate conditions as to result in the generation of a planet and a satellite of adequate mass to produce significant tides and stabilize the rotational axis of the planet.

Whereas it is certainly possible that rocky planets larger than Earth or Venus can form in other planetary systems, the mass of planets and satellites resulting from a collision is an important factor not only for the generation of tides, but for other aspects of the evolution of intelligent life on the mainland and, in particular, for the evolution of trees and arboreal animals.

Trees and hands

The reason why this is important is that, as we discussed in Chapter 3, the ability of a species to change their environment and create a technology depends not only on its degree of intelligence, but also on certain anatomical conditions, in particular on the presence of extremities identical or similar to our hands. We have also indicated that the hands, and even less evolved limbs, are not common in exclusively aquatic animals and, therefore, constitute an adaptation mainly to the terrestrial environment. Recent research suggests that hands arose as a result of adaptation to arboreal life[68], that is, thanks to the existence of an ecological niche provided by trees. Only in this niche animals are under the necessary selection pressure to develop limbs adapted for grasping the branches and for the locomotion through the treetops. In our case, this selection pressure led to the development of hands. Only arboreal animals, or those which once were arboreal, appear to possess hands developed enough to be employed in manipulating the environment. In addition, the vertical position in which the arboreal animals are forced to live favored the evolution towards the bipedal posture, which frees the hands for other uses than locomotion in a non-arboreal environment. That is, although it appears that bipedalism has evolved independently at least four times during the history of life on Earth[69], only the evolution of bipedalism from arboreal life has given rise to bipedal animals with hands as sophisticated as those of primates.

Al right, but what has all this to do with the mass of the planets in planetary systems different from ours, located in the habitable zone of stars, in which life have risen, being fortunate enough to experience an original collision generating a satellite capable of causing tides?

The answer lies in the strength of gravity. A planet more massive than Earth could have higher gravitational pull on its surface if it were also more dense, that is, if its radio were not proportionally greater than that of Earth. The gravitational pull on the surface of a planetary body is directly proportional to the mass of the planet and inversely proportional to its radius. That is, not only the mass of the planets is important for the strength of gravity on their surface, but also their size. In any case, a higher gravitational pull on the surface of a planet where life could have developed and even conquered the land, would, in principle, constitute an important barrier to the evolution of arboreal animals that may develop hands.

In my opinion, two major impediments to this end would come into play on planets with superior gravitational pull than Earth. The first one would be that the greater pull would perhaps make more difficult the evolution of plants towards the different tree species. They would need stronger structures than those we now see in the trees of our planet to support its own weight. That is, the appearance of species of large trees would probably be more difficult, having to overcome a major physical barrier to their existence. It is, therefore, probable that the tree species that might appear would not be large, but only dwarf trees or bushes. In addition, trees would have to be not only strong enough to withstand its own weight, but also to withstand the weight of the animals that would live on them.

This brings us to the second obstacle for the development of arboreal animals. They would need to possess very strong and fast muscles to move properly over the treetops on a planet with greater gravitational pull than ours. If this would not necessarily constrain the development of arboreal animal species, it may prevent them from reaching a sufficient size as to allow some of them to evolve a brain complex enough to develop a superior intelligence that ' could allow the future development of a technologically advanced civilization.

It is true that bipedalism, and perhaps also the development of hands or appendages of similar sophistication, could occur in the absence of trees, but in a world without trees, maybe with only shrubs, and with a greater gravitational pull than ours, perhaps the evolution towards bipedalism were more difficult, since it would be harder to free extremities for functions other than locomotion itself. We can imagine animals with multiple limbs, which might have developed some of them as manipulative extremities; however, this has not happened on our planet with any species, perhaps with the only exception of the elephant. Therefore, in the absence of a serious selection pressure, as it is the adaptation to arboreal life, the development of manipulative extremities must not be easy from an evolutionary point of view. As mentioned, only elephants seem to have developed a fifth "limb", his trunk, which anyway does not possess, even remotely, the manipulative ability of our hands. To manipulate the environment, two of our limbs have ceased to function as supportive extremities and we had to "learn" to use only the other two to move around. On planets with greater gravitational pull than Earth it would be, in my opinion, more difficult for the animals that could have colonized the land to develop extremities not engaged in locomotion, as is our case. Moreover, the risk of loosing balance and falling down would be more important in a planet with stronger gravity, so not using all the extremities to stand and move would be dangerous. True, we can imagine that on planets with a greater gravitational pull, animals would have evolved extremities in a greater number than four, maybe six or eight, allowing them to have an adequate number of limbs to move and also to manipulate their environment. However, evolution does not work, in general, in large steps, or is directed toward a specific goal, such as "getting" hands. That is, if the ecological niche in which life develops does not encourage it in some way, evolution does not occur, or occurs much more slowly and with no specific direction. Thus, animals with more extremities on planets with greater gravity than ours would possess these extremities because they would be important for their locomotion and survival and because they would allow them to better transmit their genes to the next generation. These extremities only would become arms and hands if some species of many-legged animals were in an ecological niche in which the progressive development of their limbs into hands conferred certain advantages for their immediate survival. In the absence of large trees to climb to, it becomes more difficult. Moreover, and importantly, control of movement and proper coordination of a greater number of limbs would require a greater commitment of the brain to control this task, which may also limit the development of this organ to take over other functions, including abstract intelligence. In short, to possess superior intelligence and the ability to manipulate the world, a certain body size is needed to allow the development of a sufficiently large brain, as well as other anatomical features allowing the brain to evolve towards freeing itself, in part, from the function for which, initially, it was required, which is none other than the control of motion. If, differently from plants, animals possess a nervous system it is precisely because animals move, and locomotion needs a fine control, which is exerted by the nervous system. For this reason, extremities in excess to those we possess would require a superior involvement of the nervous system to control and coordinate their movement, which, in turn, would hinder the development of a superior intelligence.

These and other considerations that would be too long explaining here, suggest that for the evolution of life leading one day to the appearance of an intelligent species capable of developing technology, the planets on which this evolution might occur could not be of an unlimited range of sizes. At least, these considerations allow us to conclude that the size of the planets and their gravitational pull will affect the outcome and the speed of the evolution of the life forms that could develop on them, and affect the emergence of a technological civilization.

Size and gravity

But we must not forget that, according to the thesis I am defending throughout this book, for life to evolve quickly enough to reach intelligence and technology, the planets must have a satellite large enough in size to generate substantial tides and stabilize their rotational axis. A planet more massive than Earth but with a moon the same size as ours would not suffer from such strong tides, which perhaps could also slow down the conquest of land by life, or at least delay the onset of intelligence and civilization.

By contrast, a planet with less gravitational pull at its surface than ours obviously would not suffer from the impediments of a greater gravity, but it would have to possess a sufficient gravitational pull as to retain an atmosphere dense and protective enough and to hold liquid water at its surface. Mars, with only 0.38 times Earth's gravity on its surface, lacks a substantial atmosphere and cannot retain liquid water on its surface, although it held it in the past and still retains it in the form of ice caps at its poles, or in the form of glaciers at other parts of the planet[70]. This means that the mass of the planet on which a civilization could develop must be significantly greater than that of Mars, but it should not be significantly greater than that of Earth. In any case, the most important factor to consider is the ratio between the masses of the planets and satellites resulting from the collisions, which would form them. It is reasonable to assume that this mass ratio should lay within a certain range of values, which may be more or less broad, but always limited. In our case, the Moon has only 0.0123 times the mass of Earth, but it exerts a significant gravitational effect on our planet[71]. Clearly, satellites greater than the Moon would exert a greater influence, but smaller satellites could not exert enough gravitational effect. Similarly, the mass of the greater planet of the double system, as we have mentioned, should be great enough to retain an atmosphere, lacking which life would be impossible, since water would be lost in space. In addition, a small mass in relation to its satellite could soon lead to the occurrence of the so-called tidal lock. This is the phenomenon by which the Moon rotates on itself with the same period it revolves around the Earth, thus always showing us the same side. The small size of the Moon in relation to that of Earth caused that the tidal lock of the Moon with the Earth was reached only a few thousand years after their formation. This means that if the mass and gravity of the planet were not much larger than those of its satellite, tidal lock would be reached much faster than in the case of the Earth, which will take still billions of years, at which point in time Earth also will show the same side to the Moon. When tidal lock is on the verge to be reached or it has been reached, tides disappear, as both planetary bodies are always showing the same side to each other. This means that if a planet would not possess enough mass to prevent a quick tidal lock with its satellite, as was the case of the Earth, tidal locking would proceed quickly tides would soon disappear and the duration of days and nights would much increase, causing a great temperature variation. In the case of Earth, more than two billion years since the origin of life were necessary until life was ready to colonize the land in the form of complex organisms. If by that time tidal lock with the Moon would have been reached, tides would had not existed, and its catalytic role on the colonization of land by life would not have taken place, in addition to the fact that midday or midnight temperature would have been too extreme to allow it.

Modern theories and models of planetary system formation around other stars reinforce the above considerations. It was found that the process of planet formation is chaotic[72], i.e., small variations in initial conditions result in enormous variations in the final state. This has resulted in the formation of planetary systems very different from ours in the size and number of planets and their distribution. Furthermore, as already mentioned at the beginning of this chapter, the global distribution of planets in our Solar System, with rocky planets orbiting close around the Sun and gas giants far away from it, seems to be the exception, rather than the rule.

The above suggests the possibility that massive gas planets revolving in the habitable zone of a star might have satellites on which life may develop. This is certainly possible, but we must remember that we are not concerned only with the development of life, but with the development of civilization. On the satellites of these giant planets that may be big enough to hold and atmosphere and liquid water, big tides could exist caused by the enormous gravity of the planets. However, the tidal epochs would last a short time on those satellites, because tidal locking would occur fairly quickly, since the locking time is inversely proportional to the product of the square of the planet's mass by the mass of the satellite[73]. Today we know that the four largest satellites of Jupiter are tidally locked with it, and Titan, Saturn's largest moon, holding an atmosphere, is also tidally locked with its planet.

This implies, again, that if life developed on satellites of giant planets revolving in the habitable zone of a star, they might not have tides for a time long enough to facilitate the colonization of land by life, in the case that life had arisen on them and sufficient land were available, despite not having suffered any collision with another protosatellite. To this difficulty others would join, even more important than the absence of tides, such as the fluctuation of their rotational axis in the absence of a stabilizing satellite, with the consequent abrupt climate changes, as well as with a potentially too long period of rotation around themselves, which would cause extreme differences of temperature between day and night, thus hampering the conquest of land by life, as mentioned above.

For the above reasons, we can prudently assume that Earth-like planets having similar satellites to the Moon, at the right distance from a central star so that life develops, colonizes the land and leads toward a superior intelligence and the development of a civilization, would be infrequent in other planetary systems. In this respect, as in many others, our planet is a privileged one.

Notes to chapter 4

1 Isaac Asimov (1971). The Tragedy of the Moon. Doubleday and Co. Editors

2 http://www.britannica.com/EBchecked/topic/595148/tide - http://en.wikipedia.org/wiki/Tides -

3 Richard Dawkins. Climbing Mount Improbable (1996) New York. Norton. ISBN 0393039307 http://arxiv.org/abs/q-bio.PE/0603034 - http://en.wikipedia.org/wiki/Fitness_landscape

4 http://www.nature.com/news/2010/100301/full/news.2010.99.html - http://en.wikipedia.org/wiki/Polar_bear

5 http://exoplanet.eu/catalog.php

6 http://kepler.nasa.gov/

7 http://www.sciam.com/article.cfm?id=habitable-planets-crowded-universe

8 Michael W. Werner and Michael A. Jura. Improbable planets. Scientific American. June 2009

9 On the formation of terrestrial planets in hot-Jupiter Systems. Martyn J. Fogg, Richard P. Nelson. Astron. Astrophys. 461:1195-1208, 2007. (http://arxiv.org/abs/astro-ph/0610314v1).

10 http://www.pbs.org/lifebeyondearth/alone/habitable.html - http://www.solstation.com/habitable.htmhttp://en.wikipedia.org/wiki/Goldilocks_phenomenon#cite_note-1

11 Kasting et al 1993, Icarus 101, 108–128 - http://en.wikipedia.org/wiki/Habitable_zone

12 http://exoplanet.eu/catalog.php

13 Elkins-Tanton, Linda T. (2006). Asteroids, Meteorites, and Comets (First ed.). New York: Chelsea House. ISBN 0-8160-5195-X. http://en.wikipedia.org/wiki/Asteroid_belt

14 http://www.nature.com/news/2009/090203/full/news.2009.78.html

15 http://www.pbs.org/wgbh/nova/tothemoon/origins.html

16 http://en.wikipedia.org/wiki/Thomas_Jefferson_Jackson_See

17 http://en.wikipedia.org/wiki/Osmond_Fisher

18 Binder, A.B. (1974). "On the origin of the Moon by rotational fission". The Moon 11 (2): 53–76. doi:10.1007/BF01877794. Stroud, Rick (2009). The Book of the Moon. Walken and Company. pp. 24–27. ISBN 0802717349.

19 http://www.answers.com/topic/douard-roche - http://en.wikipedia.org/wiki/Edouard_Roche

20 James H. Natland: Reginald Aldworth Daly (1871–1957): Eclectic Theoretician of the Earth. GSA Today, vol. 16, no. 2, 2006 - http://en.wikipedia.org/wiki/Giant_impact_hypothesis - Belbruno, E.; J. Richard Gott III (2005). "Where Did The Moon Come From?". The Astronomical Journal 129 (3): 1724–1745. doi:10.1086/427539. arXiv:astro-ph/0405372.

21 Galimov, E.M. and Krivtsov, A.M. (December 2005). "Origin of the Earth-Moon System". J. Earth Syst. Sci. 114 (6): 593–600. doi:10.1007/BF02715942

22 http://www.es.ucl.ac.uk/research/planetary/undergraduate/bugiolacchi/moonf.htm

23 CRC Handbook of Chemistry and Physics, 90th Edition. Editor(s): David R. Lide, National Institute of Standards & Technology (Retired), Gaithersburg, Maryland, USA. ISBN: 9781420090840

24 http://web.archive.org/web/20060901133923/http://www.astro.wesleyan.edu/~bill/courses/astr231/wes_only/element_abundances.pdf

25 http://www.psi.edu/projects/moon/moon.html

26 Wieczorek, M.; et al. (2006). "The constitution and structure of the lunar interior". Reviews in Mineralogy and Geochemistry 60: 221–364. doi:10.2138/rmg.2006.60.3

27 http://www.nature.com/nature/journal/v454/n7201/full/nature07047.html

28 http://www.es.ucl.ac.uk/research/planetary/undergraduate/bugiolacchi/moonf.htm

29 http://www.es.ucl.ac.uk/research/planetary/undergraduate/bugiolacchi/moonf.htm

30 http://www.spacetoday.org/SolSys/Moons/MoonsSolSys.html

31 http://nssdc.gsfc.nasa.gov/planetary/factsheet/moonfact.html

32 http://www.es.ucl.ac.uk/research/planetary/undergraduate/bugiolacchi/moonf.htm

33 http://www.esa.int/esaMI/Operations/SEMM17XJD1E_0.html

34 http://en.wikipedia.org/wiki/Giant_impact_hypothesis - Where Did the Moon Come From? Edward Belbruno et al 2005 The Astronomical Journal 129 1724-1745. doi: 10.1086/427539

35 Robin M. Canup. Simulations of a late lunar-forming impact. Icarus 168 (2004) 433.

36 http://www.boulder.swri.edu/~robin/moonimpact/

37 http://articles.adsabs.harvard.edu/full/2003A%26G....44b..22S

38 Cohen, B. A.; Swindle, T. D.; Kring, D. A. (2000), "Support for the Lunar Cataclysm Hypothesis from Lunar Meteorite Impact Melt Ages", Science 290 (5497): 1754–1755, doi:10.1126/science.290.5497.1754

39 Icarus (10.1016/j.icarus.2009.07.015)

40 Scott, C; Lyons, T. W.; Bekker, A.; Shen, Y.; Poulton, S. W.; Chu, X.; Anbar, A. D. (2008). "Tracing the stepwise oxygenation of the Proterozoic ocean". Nature 452 (7186): 456–460. doi:10.1038/nature06811

41 Hays, J.D.; Imbrie, J.; Shackleton, N.J. (1976). "Variations in the Earth's Orbit: Pacemaker of the Ice Ages". Science 194 (4270): 1121–1132. doi:10.1126/science.194.4270.1121 - http://www.ncdc.noaa.gov/paleo/milankovitch.html

42 http://en.wikipedia.org/wiki/Albedo#Snow

43 http://www.thermexcel.com/english/tables/eau_atm.htm - http://www.engineeringtoolbox.com/specific-heat-fluids-d_151.html

44 Richard Lathe (2004). Fast tidal cycling and the origin of life. Icarus 168(1), 18-22.

45 http://www.asi.org/adb/m/03/05/average-temperatures.html

46 http://en.wikipedia.org/wiki/Tidal_locking#Final_configuration

47 http://www.daviddarling.info/encyclopedia/C/COROT.html

48 http://www.daviddarling.info/encyclopedia/G/gravlock.html

49 http://www.daviddarling.info/encyclopedia/H/habzone.html

50 http://www.astro.cornell.edu/academics/courses/astro101/herter/java/evolve/evolve.htm - http://en.wikipedia.org/wiki/Stellar_evolution

51 http://www.nasa.gov/vision/universe/newworlds/HabStars.html

52 Clark R. Chapman & David Morrison (January 6, 1994), "Impacts on the Earth by asteroids and comets: assessing the hazard", Nature 367: 33–40, doi:10.1038/367033a0

53 http://www.nature.com/nature/journal/v420/n6913/full/nature01238.html

54 http://www.spaceref.com/news/viewpr.html?pid=8834

55 http://blogs.discovermagazine.com/badastronomy/2009/10/27/asteroid-exploded-over-indonesia-weeks-ago/

56 http://www-th.bo.infn.it/tunguska/

57 Bostrom, Nick (2002). "Existential Risks: Analyzing Human Extinction Scenarios and Related Hazards". Journal of Evolution and Technology 9. http://www.nickbostrom.com/existential/risks.html

58 http://en.wikipedia.org/wiki/Cretaceous–Tertiary_extinction_event

59 Petro, Noah E.; Pieters, Carle M. (2004-05-05), "Surviving the heavy bombardment: Ancient material at the surface of South Pole-Aitken Basin", Journal of Geophysical Research 109

60 http://www.nature.com/nature/journal/v453/n7199/full/nature07070.html

61 Schultz, Richard A.; Frey, Herbert V. (1990). "A new survey of multi-ring impact basins on Mars". Journal of Geophysical Research 95: 14175–14189. doi:10.1029/JB095iB09p14175. - http://planetarynames.wr.usgs.gov/jsp/FeatureNameDetail.jsp?feature=62601

62 Thomas R. Watters, James W. Head, Sean C. Solomon, Mark S. Robinson, Clark R. Chapman, Brett W. Denevi, Caleb I. Fassett, Scott L. Murchie, Robert G. Strom (2009). Evolution of the Rembrandt Impact Basin on Mercury. Science, Vol. 324 (5927) pp. 618–621.

63 Douglas N. C. Lin (May 2008). "The Genesis of Planets" (fee required). Scientific American 298 (5): 50–59. doi:10.1038/scientificamerican0508-50

64 http://www.skyandtelescope.com/news/home/4353026.html;

http://www.scientificamerican.com/article.cfm?id=double-impact-may-explain&ref=sciam

65 Douglas N. C. Lin (May 2008). "The Genesis of Planets". Scientific American 298 (5): 50–59.

66http://www.sciencenews.org/view/generic/id/46658/title/Extrasolar_planets_at_full_tilt

67 *http://www.nature.com/nature/journal/v470/n7332/full/nature09760.html*

68 http://www.ncbi.nlm.nih.gov/pubmed/19667206. Kivell TL, Schmitt D. (2009). Independent evolution of knuckle-walking in African apes shows that humans did not evolve from a knuckle-walking ancestor. Proc Natl Acad Sci U S A. 2009 Aug 25;106 (34):14241-6. Epub 2009 Aug 10.

69 http://www.philosophistry.com/static/bipedalism.html

70 Lodders, Katharina; Fegley, Bruce (1998). The planetary scientist's companion. Oxford University Press US. p. 190. ISBN 0195116941.

71 Wieczorek, M.; et al. (2006). "The constitution and structure of the lunar interior". Reviews in Mineralogy and Geochemistry 60: 221–364. doi:10.2138/rmg.2006.60.3.

72 Douglas N. C. Lin (May 2008). "The Genesis of Planets". Scientific American 298 (5): 50–59.

73 B. Gladman et al. (1996). "Synchronous Locking of Tidally Evolving Satellites". Icarus 122: 166. doi:10.1006/icar.1996.0117

Chapter 5: Civilizations

The origin of the Moon as a result of a major collision under particular conditions makes it very unlikely that other Earth-like planets may possess similar satellites, large enough, in any case, to cause tides over billions of years, stabilize the rotational axis of the planet and generate a quick spinning. For this reason, it might be also unlikely that other technologically advanced civilizations exist, at least in our galaxy; and more so when taking into consideration that the Moon, after helping the emergence of civilization, has also greatly enhanced the advancement of technology.

The Moon has significantly enhanced the advancement of technology, of course. Due to its proximity to Earth, the Moon has stimulated the beginning of space exploration, not only by robots and space probes, but also by humans. Space exploration has allowed the development of numerous technologies that many of us enjoy today. Had not the Moon be so close to Earth, we would have not even dreamed of sending a human being to outer space. Not even with the technology available today, after sending several men to the Moon, do we feel sure enough about our capacity to sending humans to Mars. Therefore, we would be much less assured if we had not sent somebody to the Moon, which certainly would have happened if it did not exist (and despite of this, although I do not believe it, our species and civilization would have appeared on Earth).

For all these reasons, it is likely that even if other civilizations on other planets similar to Earth exist, they are not as developed as ours, even if ours may appear primitive to some. But, can we estimate how many civilizations might exist in the universe, or at least in our galaxy?

The Drake Equation

Yes, we can! We can since, in 1960, Dr. Frank Drake[1] (now Professor Emeritus of Astronomy and Astrophysics at the University of California) published his famous equation that allows performing this estimation according to a number of factors. The values of these factors are still uncertain, and how many technologically advanced civilizations may exist in the universe depends on how we estimate them. What is this equation and what are these factors?

Before we dive into this equation, let me ask for your help to dispel the "evil spirit of anti mathematics". This spirit is the one that gets to you to leave a book, or run away from continue reading it, at the first hint that it contains a mathematical equation or a mathematical formula. The Drake Equation, as all equations, is a mathematical formula, but it is very simple. We are going to explain it very clearly and you will see how the "evil spirit of antimatemathics", which is a very unscientific spirit, as all spirits good or evil are, will leave us alone this time.

The Drake equation [2] tells us that the number of civilizations in the galaxy, which we will call N, is equal to the product of several factors:

$$N = N^{*}.f_p.n_e.f_l.f_i.f_c.f_L$$

Let's go now on explaining slowly and clearly what all these factors mean.

N is the number of civilizations in the galaxy, or the entire universe (depending on what we want to estimate), that are in a stage of development advanced enough as to be able to communicate with us. As the civilizations we could more easily come into contact with will dwell in our galaxy, we will estimate only this number. What the Drake equation tell us, once again, is that the number of advanced civilizations is calculated by multiplying a series of numbers, represented by the letters on the right side of the equation. What values represent each of these letters?

$N*$ (N-star) is the number of stars in the galaxy. It is indisputable that the number of civilizations that can develop in any galaxy depends on the number of stars it contains, or it has contained, throughout its existence (some stars die and others are born from the remains of them, as happened to our Sun).

A galaxy with a greater number of stars may also contain more stars with planets orbiting around them. In any case, it is likely that only a fraction of stars that now exist or have existed in the galaxy possess orbiting planets. This fraction is represented by the number f_p, which means nothing else but the fraction of stars with planets (fraction of planets, f_p). It is equally undeniable that life could emerge only on the surface of planets.

However, not all planets have the conditions necessary to sustain life. To our knowledge, only one planet in the Solar System possesses them: ours. The fraction of planets per star capable of sustaining life is represented by the number n_e.

Probably, however, life will not emerge on all the planets capable of sustain it. The fraction of planets capable of supporting life on which life actually arises is represented by the number f_l. (Fraction life, life fraction).

However, not on all planets on which life arises, intelligence will emerge too. The fraction of planets on which life appears, after which intelligence also appears, is represented by the factor f_i (fraction intelligence, intelligence fraction).

Again, not on all the planets on which an intelligent species has appeared it will evolve to develop a technological civilization, capable of communicating with others through space. This will only happen in a fraction of planets having intelligent life forms. This fraction is called f_c (fraction civilization).

Finally, once appeared, a technologically advanced civilization will not last forever. It is reasonable to think that civilizations, like biological species, have certain longevity and will disappear after some time. This time is denoted by f_L (fraction Longevity), which is the amount of time a civilization delivers energy into space with the capacity to be detected by others, or the time that a technological civilization can travel through space in search for other civilizations.

How many civilizations are estimated to exist in the galaxy by applying the Drake equation? Obviously, the estimated number of civilizations depends on the value we place on each of its factors. For this reason, the estimated number of civilizations fluctuates greatly according to the optimism or the pessimism of the different authors. This number varies from 0.05 (we are alone) to more than 10,000 (abundant).

The Moon factor

If we recall the Fermi paradox, explained in Chapter 2, that is, if it is true that if there were numerous civilizations in the galaxy some should have contacted with us already, then obviously there are not many civilizations in our galaxy capable of an adequate technology to communicate with others through space. Thus, the number of civilizations in the galaxy must not be very large, certainly fewer than 10,000.

The reason for the existence of few civilizations, according to the Drake equation, could be explained just by one of its factors being extremely small. For example, if the factor n_e, i.e. the fraction of planets per star capable of sustaining life, were very small, this alone would explain the absence of other civilizations in our galaxy. Of course, if not only one of these factors, but two or more, are of a small value, the number of civilizations in the galaxy would be further reduced.

If the arguments and considerations on the Moon that we have discussed in the previous chapters are true, we can assume then that technologically advanced civilizations will only develop, quickly enough at least, on double planets with particular features, similar to those found in the Earth-Moon system. That is, we can assume that at present time on our Earth, but in the past for other stars of the galaxy, a past proportional to the distance in light-years they are from us, civilizations with the ability to communicate with others have developed only on double planets similar to the Earth-Moon system. Then, the existence of double planets around a star would be a prerequisite for the development of a civilization able to contact with us now, although obviously civilizations will not always develop with certainty on these planets.

If these considerations are true (at least they seem reasonable to me) it is necessary to introduce a new factor, of my invention, in the Drake Equation. I like to call this the Moon factor, f_m. The Drake equation would then be amended as follows:

$$N = N^*.f_p.n_e.f_l.\mathbf{f_m}.f_i.f_c.f_L$$

The factor fm placed in that position means the number of planets on which life has evolved that possess a satellite large enough to exert a substantial gravitational influence on them. Only on these planets intelligence could evolve quickly enough to develop a technological civilization, although it will not develop on all of them (for instance, as discussed in Chapter 3, not on those completely covered with water). This fraction would be given, as stated, by the factor f_i, which for that reason should be placed behind factor f_m. Clearly, the f_m factor has a very small value, further reducing the product of all the other factors of the Drake equation. This new factor would decrease the estimated quantity of civilizations in the galaxy below the most pessimistic values. It not only would entail that we are alone, but that it is "a miracle" we are here. This factor could explain, therefore, the reason behind the Fermi paradox and resolve, in fact, this paradox.

The Great Filter

In this regard, we must mention that a theory, called the Great Filter[3], has been proposed to explain the Fermi paradox. This theory claims that since civilizations are very few in the galaxy (perhaps only ours, which is the only one we know, in fact, for the time being) there is a Great Filter that prevents their development. The Great Filter is hard to overcome, so only a few civilizations succeed in doing it. This filter may be composed of barriers to the development of life, intelligence, or civilization, or could consist of circumstances leading to an early death of technological civilizations once they have emerged.

Let us pause for a moment on this last point. These circumstances, if any, cannot be based on factors outside the civilizations themselves. To explain the scarcity of other civilizations in the galaxy we cannot presume, for example, that asteroid collisions with their home planets have originated their extinction. Although this might happen to some civilizations, it is unlikely it would happen to all, or even to most of them. Thus, the Great Filter, if is to be faced in our future, should emerge as an effect of the development and evolution of civilizations themselves, i.e., it must be a factor intrinsic to them. It is true, at least it appears true to me, that the fragility of our technological civilization seems to increase as technology advances. Before the discovery of nuclear energy and atomic bombs, the risk of destructing our own planet, or at least humanity and human civilization, was much lower than after that discovery. Other technologies also pose global risks that could cause our destruction as a civilization, or even the extinction of the human species. Indiscriminate and irrational use of planetary resources, caused by the technological development itself, is also a threat to the sustainability of life, of biodiversity, and, consequently, of our technological civilization, as we are currently experiencing.

Moreover, the increase in specialized knowledge, which is necessary to the proper functioning and progress of civilization, entails that a particular knowledge is possessed by fewer and fewer members of society. In this sense, the social system under which knowledge is accessed and employed, which is not only limited to availability in the Internet or other media, but implies also the know-how, the practical knowledge, is becoming less robust in the sense which defines the robustness of a system, which is proportional to the redundancy of the system. That is, a robust system possesses redundant functioning mechanisms, allowing replacement in the case one or more of them fail. If fewer and fewer people know how to perform a task or make something, and if these people disappeared or their numbers were greatly diminished, for example as a consequence of war or an epidemic, whatever the specialized knowledge they had would be lost with them, or could not be implemented properly any further. It is in this sense in which I claim that the advancement of knowledge makes a technological civilization less and less robust, that is to say, more and more fragile. How many people worldwide know how to control nuclear power or make state of the art TV sets, or keep the Internet working properly, or produce automobiles, or modify the DNA to produce transgenic organisms ...? Only a tiny fraction of the world population knows how to do these things. If that fraction were seriously impaired, perhaps by a natural disaster (e.g., collision with an asteroid) or by an artificial one (nuclear war), the ability of civilization to continue using and developing a given technology might be also impaired.

Thus, as knowledge advances and accumulates, and technology becomes more and more sophisticated, civilization becomes more and more fragile. If this is typical of all technological civilizations that may exist or have existed in the universe, and it well might be, it is possible that in the future our civilization reaches a degree of fragility causing its destruction. Furthermore, since the advance of technology makes us ever more dependent on it, it is possible that if a given advanced technology is lost, this causes a domino effect throughout civilization, because it would not be always possible to recover less advanced technologies to substitute for the lost one. In fact, the knowledge of older technologies that could help alleviate the situation could also be lost, precisely because the more advanced ones superseded them, and recover such knowledge may be a slow and difficult endeavor. In addition, our dependence on certain technologies currently required to run our civilization may be too great and cause enormous damage if they get lost. To give you an idea, think about what many of us would become if the Internet disappeared tomorrow. Moreover, the actual disappearance of a civilization due to a particular fragility is not necessary. From the point of view of cosmic loneliness, it would be sufficient if the fragility just caused an inability to communicate with other civilizations, or an inability to travel through outer space. That is, it would suffice that a particular problem converted civilizations, wherever they were in the universe, into forced autistic.

It is therefore possible that the Great Filter be very near in our future and may cause our destruction, or the regression to a less technologically advanced point from which we may advance only to regress again, in a cyclic, and tragic, repetition. The reason why the Great Filter should be close (though close may mean three, four or even more centuries) is that if the longevity of civilizations were high, i.e. the value of the fL factor in the Drake equation were elevated, not finding a filter particularly strict in the other factors, as we have assumed, then civilizations would be numerous and we could have detected some, or someone could have made contact with us. Since this is not the case, this would imply that the downfall of our civilization, or at least our technological regression, would be near, as it would have happened to other civilizations in the galaxy before ours.

However, it is also possible that the Great Filter is in our past, and that ours is one of the few civilizations, perhaps the only one in the galaxy, which has overcome it. There are numerous possibilities compatible with the Great Filter being in our past. For example, the Great Filter could be the emergence of the first molecule able to reproduce itself. Without reproduction, evolution, and obviously life, is not possible, because only what reproduces evolves.

It could also be that the Great Filter was the development of photosynthesis, the process necessary to generate the atmospheric oxygen allowing today's aerobic life processes. These processes are necessary to generate the large amount of metabolic energy needed by complex live organisms. Atmospheric oxygen also allows the formation of the ozone layer that protects land organisms from ultraviolet radiation, allowing them to live outside the aqueous medium, a layer which also had played an important role on the conquest of land by life and, therefore, on the development of intelligence[4]. An anaerobic world probably would not be able to generate the amount of energy resources needed for the development of complex organisms, the only ones able to reach a level of intelligence necessary to allow the development of a civilization.

Another candidate for a Great Filter may be the generation of eukaryotic organisms, which took about two billion years on our planet, as we said, and it is believed to have happened only once in the history of life[5]. Only eukaryotes can generate enough metabolic energy to maintain large and complex genomes, with the genes necessary for the future evolution of multicellular organisms, which in our planet are made up exclusively of eukaryotic cells.

Anyway, it is not my intention to analyze an exhaustive list of possible Great Filters in the past, all of which have been overcome by life on our planet; otherwise we would not be talking about it. My intention is, however, to propose that the most likely and more limiting Great Filter lays in our past, not in our future, and that this Great Filter is the generation of double planets with a ratio of masses into a given range and located at the habitable zones around the stars, i.e. those in which the temperature is compatible with the existence of liquid water. This does not mean that the origin of life, eukaryotic organisms, or photosynthesis is easy and always happens on any planet located at the habitable zone of a star. Perhaps these processes are part of a big global Great Filter composed of multiple limiting factors that prevent the galaxy from being teeming with intelligent beings and civilizations. But, as already discussed, the presence of a substantially sized satellite orbiting around a planet located at the habitable zone of a star is a very rare event, because it probably can only occur after a last collision happened within a very limited range of parameters. If, as I claim, the presence of a large moon orbiting a planet is necessary for the development of intelligent organisms that can evolve towards the generation of a technological civilization, the formation of a satellite like our Moon would then constitute the Great Filter, or rather, the Great Help, the driving impulse that our civilization had to appear on Earth, and do so earlier, perhaps, than any other in the galaxy. Of course, to ensure that this conclusion is correct we must await future data on the presence of life on other planets, which hopefully we will get soon. If life were common on other moonless planets located at the habitable zones of stars, but none had intelligent enough organisms, this would support the hypothesis I have been proposing.

In any case, as humanity waits for these data, the likelihood that we are alone in the galaxy, for the reasons explained before, is not small. And if we were truly alone, this would point towards a bright future for humanity, if we learn to manage it wisely. We will talk about this in the next chapter, which also, paradoxically, will help us understand why it is very likely that, in fact, we are alone in the galaxy.

Notes to chapter 5

1 http://www.seti.org/Page.aspx?pid=418

http://en.wikipedia.org/wiki/Frank_Drake

2 http://www.dbskeptic.com/2009/04/19/the-drake-equation/

http://en.wikipedia.org/wiki/Drake_equation

3 http://hanson.gmu.edu/greatfilter.html

4 http://www.nas.nasa.gov/About/Education/Ozone/ozone.html

5 Pisani D, Cotton JA, McInerney JO (2007). "Supertrees disentangle the chimerical origin of eukaryotic genomes". Mol Biol Evol. 24 (8): 1752–60. doi:10.1093/molbev/msm095. PMID 17504772.

Chapter 6. Futures

So far, we have walked a path based on known scientific data to conclude that it is not likely we, earthlings, are accompanied by other intelligent civilizations near us to which we can communicate. In this chapter, however, we are going now to enter an area of speculation about other potential universes and about the future of the human species in our universe. The mind trek I propose the daring readers might be somewhat rough for weak spirits. However, I advise you to mind travel with me because, at the end of the trip, we may perhaps encounter some hope for us and for the universe.

The anthropic principle

In my opinion, one of the most startling scientific discoveries has been establishing that if the laws and constants of Nature were even slightly different than they currently are probably we would not be here to talk about them, because our existence would be impossible. This has been called the "Anthropic Principle". We have already discussed this briefly in Chapter 1, but we should explore it more in depth here.

The degree in which the laws of Nature may be modified so that they continue allowing our existence is very small. For example, if the mass of the proton were to increase by only 0.2%, the rest of the laws and constants remaining the same, building a single atom heavier than hydrogen would become impossible, and life would be, therefore, also impossible. Similarly, certain atoms, in particular carbon, would not exist in a universe in which nuclear forces were just subtly different from ours.

The proponents of the Anthropic Principle in its most strict version claim that the laws and constants of the universe are as they are, and the universe is now as it is, *because* humans are here to observe it and try to understand it. That is, our existence and that of other possible self-aware observers of the universe is the ultimate cause, according to Aristotle's[1] definition, that the universe is the way it is at present.

However, recent discoveries indicate that the universe is an integrated system of parts and laws, like a machine. If we modify a part to a machine that is working well, most likely it will stop working. But if, at the same time we modify a part, we also modify others to compensate for the difference, the machine may continue to operate perfectly well.

With this idea, theoretical physicists Alexander Jenkins and Gilad Perez have studied whether universes different from ours, not just in one physical law or constant, but in many of them, could evolve towards complexity and life[2]. Their findings indicate that this is indeed possible and, therefore, that our universe is not as special and unique as it might seem. The findings also indicate that in the case parallel universes different from ours were formed or are still forming in other Big Bangs, some of them may also harbor life, and (why not?) intelligent beings capable of developing a technological civilization.

However, the potential theoretical universes capable of supporting life are a small minority compared to the possible universes that would not allow it. It is quite right to think that if our universe were very different than it is, we could have either not arisen or not survived in it, much less have developed a technological civilization. Moreover, the possible differences between universes are not only confined to the laws of Nature, and they include other contingencies, such as the age. If, being the laws of Nature the same, the universe had only 10% or 15% of the current age, probably it would not have produced enough heavy chemical elements in the heart of stars that die as supernovae as to allow the formation of rocky planets, like Earth, where life could develop. And if the universe were ten times older, all the stars would have died, or would be in the stage of white or brown dwarfs, and stable planetary systems would no longer exist [3] . Life and consciousness would be also impossible.

Moreover, some laws of Nature cannot be very different from the current ones, in any possible universe, should they support life. For example, if the force of gravity were much stronger than it is, the expansion of the universe after the Big Bang could have not occurred with the extent to which it has happened up to now, and probably the universe would have quickly collapsed back to the singularity from which it is assumed to have originated at the Big Bang, with no time to generate planets and life. And if, on the contrary, the force of gravity were much smaller, not enough matter might have been accumulated into stars and galaxies after the Big Bang, and the generation of elements heavier than hydrogen and helium, like carbon, the basis of life, could not have happened. Life, again, would be impossible.

The Anthropic Principle seems, therefore, to provide an answer to the following question: why physical laws are how they are, and physical constants have those values and no others? Although the assortment of possible laws and constants is not limited to those of our universe, the answer is that if they were different we would not be here to discover and study them.

However, apart from other considerations, there are two main positions on this issue. These positions are summarized in, one: the universe is the way it is because otherwise we would not be here to observe it (this seems a simple truism, better technically called tautology, and it is called the Weak Anthropic Principle); two: the universe and its laws are the way they are so that we can be here observing it. That is, it is *designed* to allow and lead to our existence. Clearly this latter idea is emotionally more satisfying for us than the first one, but this does not make it more rational, or closer to the truth.

The first interpretation of the Anthropic Principle supposes that, because existing things must be some way or another, our universe is just the way that allows the emergence of intelligent observers inside it. But this particular "personality" of the universe, even if it allows our existence, is not necessarily special in any way, because any other different universe, even if it did not allow the appearance of intelligent beings in it, would also be particular in some way. It would be that one, not another. Since any existing universe must be in a given way, we cannot conclude that ours is more "special" than any other, although its particularities have allowed us to be here talking about it. Other possible universes would possess other characteristics (for example, they might contain more or less different chemical elements), but we should not conclude for that reason that they are in that way *to make such particularities possible.*

However, the second interpretation of the Anthropic Principle supposes that the universe is just the way it is because *it had the purpose* that intelligent observers appeared within it. For my part, I would like to make it clear that I consider this position as a variant of the mystical-religious thought, which many physicists, plus many more non-physicists, like to embrace, given the peace of mind it provides. In this sense, some of the major religions in history maintain that humans are alone in the universe and they are, for that reason, "kings of creation". Although in this book I argue for mankind being alone in the universe, at least for all practical purposes, I do so from a scientific standpoint, not from belief. I certainly do not believe that the universe was created and designed by a God. This would leave open the important question of what is the origin of this God, who designed him or her, and what is the purpose of his or her existence, which cannot be to give a meaning to mine. These problems would join the question of how a God so powerful, if benevolent, let bad things happen in this world.

Aside from religious considerations, from a scientific standpoint, it is clear that if the universe is the way it is because *it has the purpose that intelligent observers appear within it*, it is not very sensible to think that it is that way so that we, humans, are the only intelligent observers. Anyway, the universe is also in such a way that allows the formation of billions of billions of stars and planets. We can, therefore, assume that the universe is designed so that a myriad of stars and planets can be generated within it, rather than matter

179

being dispersed in huge gas nebulae, allowing, perhaps, the formation of only a single star and a single planet. It is sensible, therefore, to assume that if the universe "*wants*" intelligent beings to appear throughout their evolution, they should not be restricted to one species on a remote planet in a far-flung galaxy. On the contrary, as stars and galaxies, intelligent beings should be numerous in the universe. So in a world "*designed*" for life, intelligence and civilization, conscious observers, and civilizations created by these observers, should be abundant. Moreover, the universe should be in such a way that also allowed and facilitated communication among these civilizations, which, inevitably, would be found in various places, would possess minds with different abilities, and would allow the observation of the universe from different points of view, both physical (different observation points on different planets and galaxies) and intellectual. Communication with other hundreds or thousands of different intelligences undoubtedly would allow a better observation and understanding of the universe itself. So, if the universe is as it is *for us to be here*, as intelligent observers, so that the universe may observe and try to understand itself, we can therefore ask the question: why are we then alone as a technological species even on our own planet? Why nobody is accompanying us, and no one ever has contacted us?

If the considerations developed in the previous chapters are true, and our Moon has been necessary to enable a quick development of intelligence and civilization on Earth, it is unlikely that many other technological civilizations are around, in the galaxy and, by extension, in the universe. Furthermore, given the laws of space-time, if other civilizations existed, fluid communication among them would be virtually impossible. The universe may be of such a nature so that intelligent observers appear within it, but it is not in such a way that allows the existence of many different communicating observers, which appears a contradiction in itself.

Another problem with this interpretation of the Anthropic Principle, which, in a more or less veiled way, attempts to provide a transcendent meaning to our existence, is that it can be used to justify the existence of any product of the human culture. For example, we may state here the "Shakespearean Principle " and claim that the universe is the way it is to allow someone like Shakespeare to be born and write his works, just those that he wrote and not others. We could also state the "Sinatran Principle" the "Obaman Principle", and of course, the "Me principle", assuming similar suppositions. It is obvious that these principles are absurd. However, it is true that the character of our universe has led to Frank Sinatra singing his songs, and to Barack Obama as the first black president of the United States. But it is absurd to suppose that the universe was meant so that Sinatra sung "*I did it my way*" his way.

What is indubitably true is that Sinatra, like Obama, like Shakespeare, like you, are unique. No William Shakespeare 2 or Frank Sinatra 2 have been born, and not even cloning could make this possible, for reasons too numerous to discuss here. This seems normal with humans; we have assumed our uniqueness, for better or worse, but it does not seem so normal for life, or for intelligence. However, it may well be the case that our planet is so unique in the universe as Sinatra and Shakespeare have been on Earth. What we have discussed in previous chapters suggests so. We may be the only conscious beings observing the universe, and the only ones able to use its laws to develop a technological civilization.

What do we do now?

Let suppose by the time being that we are indeed alone in the universe, or at least in the galaxy. What do we do then? Does our existence hold a special meaning because we are alone? Is it not this what some religions have been telling us all the time, that we humans are unique, the kings of creation, and that all there is in the universe is for us to use and enjoy? Is it not all this a vulgar variant of the Anthropic Principle? It well maybe, but there is a big difference between thinking something because it suits us well, or because our elders or certain authorities told us so (religion), and thinking something because we have concluded so after centuries of studies and observations, and the emotionally disinterested application of logic (science), although the idea, or conclusion, could be the same.

In any case, it is certain that if we are alone in the galaxy, the galaxy is "ours" and, of course, we are also "ours." We must, therefore, think about what we want to do with the galaxy, with the universe and with ourselves, we must think and debate about our destiny, and we need to think about it with a great breadth of vision, projecting our thoughts to a future extending not decades or centuries, but millions of years.

First, if we are alone in the galaxy, or perhaps in the entire universe, we must learn to appreciate the extraordinarily special our planet Earth is and the special we are, as one of the few, currently perhaps the only one, technological civilization capable of exploring and understanding the universe. Both our planet and we are probably truly exceptional entities and living beings, perhaps unique in the vastness of the universe. For this reason, we must protect our planet and we must fight to solve rationally the problems of mankind. I think that in the medium to long term, the problems of humanity will be solved when we achieve a true society of knowledge and full rationality, and abandon archaic ideologies, typical of an "intellectual and technological Third World", as are some religious ideas and traditions contrary to the most elemental ethical behavior with Nature. I also believe that humanity already knows what needs to be done to end poverty, indoctrination, and the abuse of some human beings by others ... but some anti-democratic interests prevent the appropriate measures to be put in place. Perhaps the idea that we are alone, that we are unique, extraordinary, and that the universe is "waiting for us", might accelerate the necessary transformation of humanity, of which I will talk later more extensively, and make everyone understand that, above all, we must preserve our humanity and our civilization because, if it disappears, there will be no time left in the universe so that another similar may reappear, and if perhaps something had acquired a transcendent meaning thanks to our existence, it would be lost forever.

However, if as discussed in chapter one, chance does not exist, our destiny is determined however much or little we think about the future. Anyway, we are part of the natural evolution of the universe and our own evolution over time, planned or not, can perhaps come to affect its.

This idea may seem too bold, too risky. How can we get to affect the evolution of the entire universe? Admittedly, this is a bold idea. If the first organism to emerge on our planet could have looked at itself, observe the planet on which it lived, and speculate about the future, I do not think that, being wise, it would have come to the conclusion that after billions of years their descendants would change the composition of the planet's atmosphere, flooding it with oxygen, and thus dramatically affecting the evolution of life on it, allowing the emergence of multicellular animals formed by a society of billions of cells much more sophisticated than itself. However, that is what has happened. The descendants of that organism are us, in addition to other living beings, and our life has changed and continues to change the entire planet.

Something similar could happen in the case of our civilization. Our civilization proves that the existence of technological civilizations is possible in the universe, but as in the case of living beings, one civilization or another must be the first to emerge. Perhaps ours was the first one, considering the low probability of repeating similar conditions to those which have allowed our origin, some of which, but not all, have been discussed in the previous chapters. Similarly to the first bacterium giving rise to descendants that transformed the entire planet, it could be that the first civilization gives rise to descendants that will transform the universe itself. The origin of the first civilization may have implications for the future of the universe similar in magnitude to those the origin of life caused on Earth.

How? We do not know yet, not even we do know whether such a transformation is possible, but we can have fun in carrying out a speculative analysis about what might happen if our civilization, as the first microorganism, evolved and spread through the galaxy and the universe, solving forever the Fermi paradox.

Let us recall that, according to the calculations described in Chapter 1, a civilization could colonize the entire galaxy in about fifty million years, although it probably could do it in only one tenth of that time. It is thus possible, as human curiosity and desire for explorations suggest, that the fate of our civilization is to evolve, expand and conquer the galaxy and (why not?) the entire universe. This will not happen tomorrow but, if it happens, it will happen during the next millions of years, as stated above. Although it may seem a long time, it is very short considering the time left for the galaxy and the universe to "die", which for the purposes of formation of new stars and planets can be estimated in, at least, a thousand times its present age[4]. Consider that the Earth today is only 3.5 times older than when life arose on it. The universe has, therefore, a much greater longevity than that necessary for a civilization to colonize it.

Granted, this idea sounds crazy, as crazy too could have sound the idea that a simple bacterium could change planet Earth. But if one intelligent being could have analyzed the evolving capacities of bacteria, he or she might have found that possibility as likely, perhaps as inevitable. Similarly, the analysis of our possibilities for progress, as a civilization, could help convince us whether conquering the galaxy, or the entire universe, and changing its fate or its evolution, is or not possible.

Unfortunately, we cannot ask to a more advanced being than us whatever she or he might think about our chances for evolution; the first bacteria, obviously, could not do it either. But we have an advantage over the first bacteria: we have considerable knowledge about ourselves and about the laws of Nature and the universe as to venture to perform this analysis. This is what I will try very briefly below. I do not intend to be fully accurate on my predictions. I simply intend to open a window toward the future and to some of its possibilities, not on a scale of decades, but centuries, millennia, or even longer. Let's begin.

The technological singularity

For some researchers in various fields of science, humanity is close to what it has been called the *technological singularity*[5]. In mathematics, a singularity is a point in a given mathematical function that represents a break with the rest of the points of the function. It is like that the singular point no longer belonged to the function, and has abandoned it, paradoxically, as a result of the function itself. If we suppose technological progress is a function relative to time, it is clear that this function is not directly proportional, but rather exponential. For example, humanity has taken much longer to evolve from small groups of gatherers and hunters to small villages of farmers or ranchers, than to move away from these villages to the large cities of today. The progress made by mankind in the twentieth century has been indisputably greater than that made during all its previous history. And this progress, far from slowing down, seems to be speeding up.

This said, the technological singularity would suppose reaching a point where the progress of humanity makes such a leap that represents a break, not only in the history of technology, but in the evolution of humanity itself. What could this leap be? Will it happen for sure? Where is progress leading us?

To answer this question, thinkers on this subject have analyzed the factors limiting progress. Although there may be many constraints that can stop mankind's progress, there is an absolute constrain that limits it, which is nothing else than human intelligence. It is clear that mankind cannot progress towards an incomprehensible technology, although such technology may be possible, and in agreement with the laws of Nature. Indeed, only the most qualified and prepared minds are able to use some of the artifacts generated by our current technology. Consider, for instance, the sophisticated machines employed for some medical diagnosis or treatments.

For this reason, those who claim a technological singularity is possible propose that it will occur when the technological progress generates a super-intelligent machine, whose intelligence not only would be superior to human's, but it will be aimed to its progressive self-increase along new generations of machines designed by the very super intelligent machines themselves, which will be able to self-replicate too[6]. In the words of the English mathematician John Good Irvin (1916-2009), who proposed the concept of technological singularity, at that moment humanity would reach a point of "intelligence explosion"[7]. Of course, this intelligent machine would be the last invention of mankind, because once it is made, human beings would be left behind; they would be transcended, overcome in evolution by the emergence of a new species: the machine with progressive super intelligence.

It is convenient here to open a parenthesis and mention something I consider important. In chapter 2 we said that life elsewhere in the universe must be based on the chemistry of carbon. However, it is possible that other "live" beings based on another kind of chemistry, for example silicon and iron, could evolve from carbon-based living beings. This could happen in our planet, although I do not consider it very likely. However, if this had happened in some other planets of our galaxy in which a technological civilization could have developed, it is quite likely that these super-intelligent self-replicating beings would not have the same kind of problems than water and carbon-based beings have to travel through outer space and colonize the galaxy. Therefore, if a super-intelligent machine were the next natural stage in the evolution of intelligent beings, the fact that extraterrestrial civilizations have not contacted us would be further evidence in favor that the resolution of the Fermi paradox resides in that we are alone, or at least we are very rare in our galaxy.

In any case, in view of our current progress, only possible by the use of the most intelligent machines we have: computers, there may be some truth to the idea that a machine of ever increasing intelligence, self-designed and self-improved, is possible. Without today's computers, used to design newer and faster computers and more and more advanced electronic components to make even more powerful computers or other increasingly sophisticated machines, the speed of progress as we know it would be impossible.

But the idea that machines may supersede entirely the human race has been, perhaps, too bold for some proponents of the technological singularity. They, however, suggest that the technological singularity will happen through the evolution of the human species, aided by intelligent machines, but not so clever, or mean, as to displace humanity. In this scenario, human beings will become transhuman, or posthuman[8], in any case more advanced and sophisticated beings than current humans.

The technological singularity has been also compared to what is called a meta-evolution, i.e., an evolution to a superior system. This development represents a change in the level of organization or control, and the emergence of a new organized system from simpler elements or systems[9]. An example is the transition from unicellular to multicellular organisms; another one is the appearance of individuals with defined roles and reproductive castes, as in social insects.

However, the idea of the technological singularity, like any idea issued to try predicting future trends, has its detractors. Among them we encounter eminent scientists, like Marvin Minsky[10], or Gordon Moore[11], the proponent of Moore's Law, which says that computing power doubles every two years so that their progress is exponential. In spite of this, Moore does not believe that the technological singularity would ever really happen.

Improving the human race

For my part, I do not believe either that a technological singularity generating a break point in the evolution of humanity would happen. However, I do believe we have a wide margin to evolve driven by technology, thanks to the scientific and technological knowledge we already have, which we will undoubtedly increase in the not too distant future. What I think will happen in the coming years is that there will be an increasing interaction between different technologies and areas of knowledge, which will allow significant advances leading, in the medium to long term (remember we're talking, at least, about thousands or tens of thousands of years) to the generation of more intelligent human beings, with better physical qualities and, above all, better social skills and teamwork capabilities. We move towards a meta-society in which individuals will live, however, a fuller life than ours. Nevertheless, I do not believe that this progress will happen suddenly, as a result of a technological singularity in the terms explained above.

To clarify what I mean, I must mention some concepts and facts that I consider proven by science. The first one is that human beings, and in general all living beings, are machines and work according to the principles of engineering and computer science, and of course, according to the laws of Nature. What I mean by this is that living beings are very complex mechanisms that execute processes according to the inputs received and thus producing certain outputs. There is no a "life force" or a "mysterious impulse" to sustain life. The animals, and humans, have no soul, understood as a spiritual part allowing us to acquire the category of animate beings, endowed with consciousness, intelligence and free will; in short, allowing us to be free beings, if we define freedom as the ability to act on the environment independently of the laws governing matter.

In the case of human beings and animals, the control of the processes generating outputs from the inputs received depends on the structure of networks made from cells specialized in information processing: nervous system cells, whether neurons or cells supporting neuronal activity, such as glial cells. The cognitive abilities of a human being depend on the senses that receive the inputs and initiate their processing and on how these inputs are then further processed and stored in particular neural networks of our nervous systems to generate outputs. A single input, or a series of inputs, can be processed better or worse according to the neural structure involved in that processing, and according to the level of its fitness for the role this structure performs.

The structure of neural networks is determined primarily by genes and, secondly, by the change and adjustment produced by learning[12]. Without the genetic information put to work during brain development, neural structures would not be predesigned to learn how to carry out a specific function. That is, during the development of the brain, neural structures develop therein, specialized in learning and performing certain functions. For instance, neural networks are created that will be responsible for processing visual, auditory, or olfactory, information, or to learn and store the concepts proper of language.

Learning of the mother tongue, and of everything a human being can learn, like playing a musical instrument, depends on the structural adjustment caused by the learning process to specific neuronal circuits, involved in the acquisition of new concepts, which are stored in neural networks, or in the control of the behavior being learned or having been learned, whether playing the flute, or speaking Chinese. Of course, the learning ability of our brain, the ability it has to modify in a plastic way its neural circuits, depends not only on the experiences we receive from the external environment, but also on our genes. I kindly remind the reader that genes are not merely repositories of information, but within living cells they are the manufacturers of the parts allowing the cellular machinery to operate. In the case of neurons, genes manufacture the parts allowing the connection with other neurons, the modification of existing connections, the formation of new connections or the destruction of unneeded connections.

This implies that, if we possessed sufficient knowledge, we could genetically engineer specialized neural structures to learn how to perform cognitive functions that humans lack, as for example to count instantly and accurately the people who are in a cinema or auditorium and estimate their approximate average age, as well as other more useful cognitive prowess. On the other hand, knowledge about the genes involved in learning also could improve human beings in this capacity, so typically human.

As the reader can see, this idea is not far from the idea of the machine that gradually improves itself with each generation by using its growing intelligence, an idea summed up in that, like all machines, human beings can be improved in the multiple subsystems that form them. These subsystems are not limited to the nervous system, of course. Human beings could be also modified to be stronger, faster, and more resilient to fatigue, in addition to make them smarter.

However, the strength, speed, and resilience will be worth little without an intelligence to properly manage them.

We should not forget that other aspects of human nature, which also depend on the structure of the brain's neural networks, also could be improved. One of them is the ability to integrate into society and to live and survive in it, working in collaboration or in competition with others. In my opinion, and in that of other scientists, our own species has survived only because it has been able to reconcile these two opposing social forces: cooperation and competition. Today, we continue being immersed in this social contradiction, and as we work closely with some, we may be competing with others. To regulate these trends, we gave ourselves huge amounts of rules and laws that attempt to perform what is called political engineering[13]. I believe our natural tendency is to increase opportunities for collaboration, and reduce the need for competition and conflict with others. However, by our very nature, we cannot help coming sooner or later into conflict and competition with others.

The knowledge of neural networks and genes that affect social behavior might allow the generation of human beings more and better integrated into society, in which they will work in cooperation, not in competition, with others. They will be human beings much less selfish and more altruistic than we are today. This will not only lead, in my utopian view, to an improved society, but also to human beings living a fuller and happier life, since for altruistic beings their own happiness lies in the happiness and welfare of others. In short, the political engineering, based on ideologies and opinions, will be replaced by social engineering[14], based on scientific knowledge, not on religious or ideological views, which often lack any scientific support.

Another aspect that can undoubtedly be improved by the technology of molecular biology and genetic manipulation is human longevity. What is known today about the causes of aging is that they are largely genetic, so it's therefore possible to think that human life can be lengthened by the molecular modification of the genes involved in the aging process. This could be achieved, I dare to imagine, with the use of biological nanomachines designed to remove the genes damaged by aging and replace or repair them to restore a younger and healthier version, which in addition to preventing aging would permit curing cancer and other genetic diseases, if they may arise. These technologies would allow the generation of longer-lived human beings, able to maintain themselves forever young in their physical and intellectual capacities, and able to quickly learn new concepts throughout their lives. No doubt, these guys will possess an enormous reservoir of knowledge and experience for society.

These ideas aimed at the improvement of human beings clash head-on with the feelings and ethical ideas that most advocate today about the good or evil, the right and wrong, of genetic and artificial improvement of our species. However, if we examine, even briefly, the evolution of the ideas held by humanity during the past centuries, and project them into the future, it is possible to come up to the realization that the ideas we consider right today, might be considered wrong in a few hundred years. For example, if we know we may give birth to a sick child, or to one with an intellectual disability, but the molecular and cellular technology required to prevent it can be applied, it would seem unethical and contrary to the moral not to use it, even if it is very expensive. Similarly, while it may seem wrong today, if we know that this or that genetic modification will ensure that our child will be smarter, healthier, stronger, and happier, in the future it may become unethical, or irrational, not to use the technological capability to perform that modification, and condemn our son or daughter to be an inferior and more unhappy human being to whom it might have been. Today, we already have biomedical technologies that most people would consider immoral not to use, such as blood transfusion. Some religions believe that blood transfusion is against the will of their god, and even when life depends on it believers refuse to receive it, or to administer it to their relatives. This refusal, based on beliefs, is considered nowadays as an aberration by most of the humankind. It is possible, therefore, that in a few centuries, perhaps, refusing to use these new biomedical technologies could be considered a moral deviation, even if this idea can disgust us today. In any case, I'm not talking here about what is right or wrong, but about what may be possible, including the fact that what we think wrong today can be considered right tomorrow. Moreover, we must consider that once our species is genetically improved, with appropriate medical and molecular monitoring, these improvements can be maintained by the traditional methods of reproduction that we all know, so it will be necessary to apply those improving technologies only over a short period of the evolution of humanity and then maintain the improvements, or continue improving very gradually.

Of course, if we could modify ourselves, we would be also able to modify other organisms for our purposes. This, in fact, is already being done today. The genomes of laboratory mice have been modified in thousands of different ways for research purposes[15]. We have changed the genomes of many plants with the intent to improve various qualities, such as its growth in high salinity lands, or the rate of fruit ripening[16]. It is possible that our technological capacity to genetically improve organisms may increase in coming years to the point where we can practically create a new organism, designed from a previous one used as a template. This could be achieved in both plants and animals. Moreover, if technology allows it, nothing would prevent us from generating thousands of organisms with different designs and select the one most suitable for our purposes, in what I call the educated application of evolution by selection. Anyway, it will be possible to generate new plants and animals, and even new specialized biological devices, resistant to certain conditions, that may help us to leave our planet and colonize nearby planets, in what would be the first step in our expansion across the galaxy.

We should not forget either that these genetic modifications may also lead to better integration with machines or to improve their use. I do not know if a human-machine integration would happen, leading to the creation of a kind of Cyborg[17], similar to those seen in science-fiction stories, like the protagonist of the popular TV series from the 70s "The six million dollars man", or the Borg, from the 90's "Star Trek: The Next Generation" TV series. However, it is possible that "biologication" of machines, for example, the making of computers based on biological processes, such as the replication of DNA, or based on artificial neural networks maintained in culture, may allow human-machine integration in a way not imagined by science-fiction, because machines would not be plastic and metal contraptions, but other biological systems similar to us.

These ideas may seem too risky, but we already have very powerful technologies, whose interrelationship will multiply exponentially the possibilities for designing new systems, new organisms, and improving human beings. One of these new technologies is synthetic biology[18], a new discipline that applies engineering principles to biological systems to generate new systems with features not found in Nature. This discipline will benefit from its interface with computer science, and in particular, with bioinformatics, a discipline aimed at the computer-assisted analysis of the molecules that are part of biological systems, and at the understanding through this analysis, of the principles governing their interactions and the operation of the molecular systems formed by them. The increasingly powerful computers and computer algorithms may allow, in a not too distant future, the design of new biological systems and even simulate how adequate they are for our purposes before synthesizing them in the laboratory or in the "molecular bioplant". These systems may be biological nanomachines designed to accomplish specific functions, such as the correction of certain mutations in certain DNA sites, removal of damaged organelles within the cells, the fight against certain pathogens, and "in situ" cellular and genetic engineering.

It is also necessary to take into consideration the development of nanotechnology[19], which aims to build new systems and devices at the nanometer scale. To give us an idea of this scale, we need to consider that a nanometer is a billionth of a meter, which corresponds, more or less, to the relationship between a marble and the size of the Earth. The diameter of a strand of DNA is only two nanometers. The construction of nanodevices involves the manipulation of atoms almost individually. The development of this technology, together with those mentioned briefly above, could lead to the design of new mechanisms that could be integrated with biological organisms, like those developed by the technology of synthetic biology.

The brief overview of the technologies future generations await supports the idea that other technological civilizations, having arisen only million years before us elsewhere in our galaxy, which is not much time to a galactic scale, could have already developed them and would have perhaps completed their galactic colonization, as we said at the beginning of this book. All these technological possibilities support the idea that space travel is possible, either for living beings modified to do so, or for the machines they could create. However, it seems we are alone, suggesting that there are not many advanced civilizations in our galaxy, if there is any in addition to ours. In any case, if despite this there were many civilizations in the galaxy, our loneliness indicate that they must have appeared in a period too close to that of our own appearance, because otherwise we would know about them, thanks, among other things, to the SETI program.

In conclusion, the possible development of new technologies already available today, albeit in a rudimentary way, and the interrelationship that will develop among all of them, allows us to imagine that substantial modifications of human beings, aimed at increasing their physical and intellectual abilities, are possible in the coming centuries. These improvements would allow the generation of humans much better adapted to live in outer space or on other planets of the Solar System, as a first step, and later, perhaps thousands of years later, to live on extra solar planets. At that time scale, much greater than what usually concerns us as the future, the beginning of the expansion of humanity to other planets in the galaxy may be possible. Hopefully sooner, thanks also to the use of scientific knowledge and reason, mankind will get rid of the last strongholds of obscurantism and irrationality, will be able to control the emotions that try to justify and motivate all kinds of unfounded ideologies and tribalism, from which we still suffer today, and will progress steadily on the path of rationality to decide their own destiny and, perhaps, the destiny of the entire universe, to which we can bring life and civilization.

Notes to chapter 6

1 Aristotle. Aristotle in 23 Volumes, Vols.17, 18, translated by Hugh Tredennick. Cambridge, MA, Harvard University Press; London, William Heinemann Ltd. 1933, 1989.

2 *Alexander Jenkins and Gilad Perez. Looking for life in the Multiverse. Scientific American, January, 2010. pp 28.*

3 Carter, B. (1983). "The anthropic principle and its implications for biological evolution". Philosophical Transactions of the Royal Society A310: 347–363. doi:10.1088/0264-9381/14/4/002

4 http://math.ucr.edu/home/baez/end.html - http://en.wikipedia.org/wiki/Graphical_timeline_from_Big_Bang_to_Heat_Death

5 http://www-rohan.sdsu.edu/faculty/vinge/misc/singularity.html - http://singinst.org/overview/whatisthesingularity - http://en.wikipedia.org/wiki/Technological_singularity

6 http://en.wikipedia.org/wiki/Self-replicating_machine

7 I.J. Good, "Speculations Concerning the First Ultraintelligent Machine", Advances in Computers, vol. 6, 1965.

8 http://www.transhumanism.org/index.php/WTA/constitution/

9 "Evolutionary Transitions: how do levels of complexity emerge?

10 http://en.wikipedia.org/wiki/Marvin_Minsky

11 http://en.wikipedia.org/wiki/Gordon_Moore

12 http://www.nature.com/nrn/journal/v9/n2/full/nrn2321.html

13 http://en.wikipedia.org/wiki/Political_engineering

14 http://en.wikipedia.org/wiki/Social_engineering_%28political_science%29

15 http://www.genome.gov/10005834

16 http://www.agbioworld.org/ -http://www.pnas.org/content/96/11/5937.abstract - http://en.wikipedia.org/wiki/Transgenic_plant

17 http://en.wikipedia.org/wiki/Cyborg

18 http://syntheticbiology.org/

19 http://www.crnano.org/basics.htm

THE END

www.ingramcontent.com/pod-product-compliance
Lightning Source LLC
Chambersburg PA
CBHW030006190526
45157CB00014B/540

9 781446 712450